YALE UNIVERSITY

MRS. HEPSA ELY SILLIMAN MEMORIAL LECTURES

ELEMENTARY PARTICLES

Elementary Particles

BY ENRICO FERMI

NEW HAVEN: YALE UNIVERSITY PRESS

THE SILLIMAN FOUNDATION

In the year 1883 a legacy of eighty thousand dollars was left to the President and Fellows of Yale College in the city of New Haven, to be held in trust, as a gift from her children, in memory of their beloved and honored mother, Mrs. Hepsa Ely Silliman.

On this foundation Yale College was requested and directed to establish an annual course of lectures designed to illustrate the presence and providence, the wisdom and goodness of God, as manifested in the natural and moral world. These were to be designated as the Mrs. Hepsa Ely Silliman Memorial Lectures. It was the belief of the testator that any orderly presentation of the facts of nature or history contributed to the end of this foundation more effectively than any attempt to emphasize the elements of doctrine or of creed; and he therefore provided that lectures on dogmatic or polemical theology should be excluded from the scope of this foundation, and that the subjects should be selected rather from the domains of natural science and history, giving special prominence to astronomy, chemistry, geology, and anatomy.

It was further directed that each annual course should be made the basis of a volume to form part of a series constituting a memorial to Mrs. Silliman. The memorial fund came into the possession of the Corporation of Yale University in the year 1901; and the present work constitutes the thirtieth volume published on this foundation.

Foreword

With numerous additions and subtractions the present book contains the material presented by the author in the Silliman Lectures at Yale University in April, 1950. Six lectures for the general public and six lectures for the physics students were given. In this rewriting the subject matter of the latter has been emphasized and amplified.

Many of the theoretical papers on the subject of elementary particles and of their interactions are very difficult reading except for a small, highly specialized group of theoretical physicists. This book is not written for that group. It attempts instead to make accessible to a larger number of students and, I hope, a large fraction of experimental physicists some of the most significant results of the field theories of elementary particles that can be understood, at least in a semi-quantitative way, without excessive mathematical apparatus. A reader who has followed and understood a good standard university course in quantum mechanics should not find serious difficulties in the following pages.

I have not been able to give in the text adequate references to the very extensive original literature. I am afraid, also, that in several instances I may not have succeeded in giving due credit to the originator of a particular idea. I apologize for the omissions.

Fortunately an excellent "Guide to the Literature of Elementary Particles Physics" has been prepared recently by Tiomno and Wheeler (*American Scientist, 37*, April and July, 1949), and the reader is referred to it for much detail. A list of books and review articles in which extensive additional information can be found is appended here.

H. A. Bethe, Elementary Nuclear Theory, New York, 1947.
E. Fermi, "Quantum Theory of Radiation," *Rev. Mod. Phys. 4*, 87, 1932.
W. Heitler, Quantum Theory of Radiation, Oxford Univ. Press, 1944.
E. J. Konopinski, "Beta Decay," *Rev. Mod. Phys. 15*, 209, 1943.

W. Pauli, Meson Theory, New York, 1946; "Relativistic Field Theories of Elementary Particles," *Rev. Mod. Phys. 13*, 203, 1941.

L. Rosenfeld, Nuclear Forces, New York, 1948.

J. Schwinger, "Quantum Electrodynamics," *Phys. Rev. 74*, 1439, 1948; *75*, 651, 1949; *76*, 790, 1949.

J. Tiomno and J. A. Wheeler, "Spectrum of Electrons from Meson Decay," *Rev. Mod. Phys. 21*, 144, 1949.

G. Wentzel, Quantum Theory of Fields, New York, 1949; "Recent Research in Meson Theory," *Rev. Mod. Phys. 19*, 1, 1947.

Chicago, September, 1950.

Contents

This is a collection of the notations most frequently used in this book. When the same letter is used for different meanings the various meanings are listed, sometimes with indication of the section.

a, a^* Destruction and creation operators.

a Bohr radius.

A Vector potential.

B Plane wave spinor.

c Velocity of light.

D Deuteron wave function.

e Elementary electric charge.

e Symbol for electron. Also amplitude of electron field.

e_2 Yukawa interaction constant.

e_3 Interaction constant of pion decay.

f Composite interaction constant for pion production (secs. 18, 19, 25).

g_1 Constant of the beta interaction.

g_2 Interaction constant for spontaneous muon decay.

g_3 Interaction constant for forced muon decay.

$h, \hbar$ Planck constant and Planck constant divided by 2π.

$H, \mathcal{H}$ Hamiltonian or interaction Hamiltonian.

J Current density.

$\mathcal{L}$ Lagrangian density.

m Mass of a particle, particularly electron mass.

M Mass of a nucleon.

n_s Frequently used as number of particles in state s.

N Frequently used as number of states.

N Symbol for the neutron. Also amplitude of neutron field.

$\bar{N}$ Symbol for the anti-neutron.

p Momentum

P Symbol for the proton. Also amplitude of the proton field.

$\bar{P}$ Symbol for the anti-proton.

q Relative momentum (secs. 18, 19).

Q Volume of momentum space.

r	Vector of a point in space.
s	Frequently used as index for a particle state.
T	Kinetic energy.
$u(r)$	Eigenfunction of a particle.
U	Potential of nuclear forces.
v	Velocity, sometimes speed.
V, V_0	Volumes (secs. 25, 26).
w	Energy of a particle including rest mass.
W	Total energy of the system.
W'	Energy of bombarding nucleon in laboratory system.
α	Dirac matrices (space components and vector).
β	Dirac matrix (fourth component).
γ	Symbol for a photon.
ϵ	Polarization unit vector.
$\lambda, \acute{\lambda}$	De Broglie wave length, ordinary and divided by 2π.
μ	Symbol for the muon. Also amplitude of muon field.
μ	Mass of pion.
μ_1	Mass of muon.
ν	Symbol for the neutrino. Also amplitude of neutrino field.
$\bar{\nu}$	Symbol for the anti-neutrino.
π	Symbol for pion. Also amplitude of pion field.
σ	Cross section.
σ	Pauli spin operators.
τ	Lifetime or inverse transition rate.
τ	Temperature in energy units (sec. 26).
φ	Scalar field amplitude.
ψ	Field amplitude for particles with Pauli principle.
ω	Angular frequency.
Ω	Normalization volume.

CHAPTER ONE

Quanta of a Field as Particles

1. INTRODUCTION

Perhaps the most central problem in theoretical physics during the last twenty years has been the search for a description of the elementary particles and of their interactions. The radiation theory of Dirac and the subsequent development of quantum electrodynamics form the present basis for our understanding of the electromagnetic field and its associated particles, the photons. In particular, this theory is capable of explaining the processes of creation of photons when light is emitted and of destruction of photons when light is absorbed. The field theories of other elementary particles are patterned on that of the photon. The assumption is made that for each type of elementary particle there exists an associated field of which the particles are the quanta. In addition to the electromagnetic field an electron-positron field, a nucleon field, several meson fields, etc., are also introduced.

The Maxwell equations that describe the macroscopic behavior of the electromagnetic field have been known for a long time. It is therefore natural to assume that these are the basic equations one should attempt to quantize in constructing a quantum electrodynamics. This has been done with a considerable measure of success. In the past two or three years the last remaining difficulties associated with the infinite value of the electromagnetic mass and the so-called vacuum polarization have been largely resolved through the work of Bethe, Schwinger, Tomonaga, Feynman, and others. They have been able to interpret satisfactorily the Lamb shift in the fine structure of hydrogen and the anomaly of the intrinsic magnetic moment of the electron as due to the interaction with the radiation field.

Next to the photons the particles which are best known experimentally and best understood theoretically are the electrons and positrons. In the field theory of electrons and positrons the relativistic equations of Dirac are taken as the field equations of the electron-positron field. The procedure of quantization in this case must, however, be of a type such as to yield the Pauli principle for electrons and positrons instead of the Bose-Einstein statistics that applies to the photons. This can be done with the second quantization procedure of Jordan and Wigner.

Less convincing are the attempts at a similar description of fields about which we have much scantier experimental knowledge.

Protons and neutrons which, like the electrons, obey the Pauli principle and have spin 1/2 are usually also described by a Dirac equation. This goes of course beyond our present experimental knowledge because until now no negative protons (the analogue of the positrons) have been discovered. Neither have anti-neutrons been discovered. These hypothetical particles are the counterpart of the neutron in the same sense that the positron is the counterpart of the electron. The anti-neutron differs from the neutron in that its intrinsic magnetic moment is directed parallel to the spin angular momentum, instead of anti-parallel, as it is for the ordinary neutron. Also, additional complications are encountered because according to the simple form of the Dirac theory one would expect the magnetic moment of the proton to be 1 nuclear magneton and that of the neutron to be 0. The fact that the proton has instead a moment of 2.7896 and the neutron of -1.9103 nuclear magnetons is currently attributed to the action of the meson field surrounding the nucleons. If this view is accepted we are led to the conclusion that the physical proton and neutron are in fact much more complicated objects than they seem when described in terms of the Dirac theory.

So far we have discussed particles whose basic properties are known in great detail. But there are other particles whose existence is known or suspected and whose properties are in several cases only conjectured.

The existence of the neutrino has been suggested by Pauli as an alternative to the apparent lack of conservation of energy in beta disintegrations. It is neutral. Its mass appears to be either zero or extremely small (less than a few kev energy equivalent). Its spin is believed to be 1/2; its magnetic moment either zero or very small. In the theory of the beta decay the neutrino is usually described in terms of a Dirac equation that gives two types of neutrino, neutrino proper and anti-neutrino, related to each other like the electron and the positron. This is not the only mathematical possibility. Another one has been suggested by Majorana in which there is no anti-neutrino. In the application to the beta theory worked out by Furry the Majorana theory usually gives the same results as the Dirac theory except in the case of the very improbable double-beta processes recently investigated by Fireman. The beta-ray theory based on the neutrino hypothesis has had some success in explaining the general features of the beta disintegrations and in particular the energy distribution of the emitted electrons. On the other hand, until now no really convincing form of this theory has been discovered. Instead of one satisfactory beta theory there are several of them, none quite acceptable.

A great deal of work has been devoted to the field theory of mesons first proposed by Yukawa in his attempt to explain nuclear forces. The meson of Yukawa should be identified with the π-meson of Powell (briefly called here pion). The μ-meson of Powell (called here muon) is instead a disintegration product of the pion, only weakly linked to the nucleons and therefore of little importance in the explanation of nuclear forces. The Yukawa theory has proved a very valuable guide in experimental research and probably contains many correct leads to a future theory. In particular, it is partly responsible for the discovery of the production of mesons in the collision of fast nucleons. On the other hand, the attempts to put this theory in a quantitative form have had very mediocre success. Often a ponderous mathematical apparatus is used in deriving results that are no better than could be obtained by a sketchy computation of orders of magnitude. This unsatisfactory situation will perhaps improve only when more experimental information becomes available to point the way to a correct understanding.

The purpose of this discussion is not to attempt a mathematical treatment of the field theories but rather to exemplify semi-quantitative procedures that are simple and may be helpful in the interpretation of experiments. There are several cases in which not much would be gained by a more elaborate mathematical treatment since a convincing treatment has not yet been discovered. In other cases the qualitative discussion presented here may serve as an introduction to more complete elaborations of the subject.

2. THE ELECTROMAGNETIC FIELD

As a first example of the quantization of a field the case of the electromagnetic field and of its photons will be discussed. Unfortunately the electromagnetic field has a rather complicated structure since it is specified at each point by the electric and the magnetic vectors. On the other hand, it is the most familiar field and its quantum properties are most clearly understood.

In quantum mechanics observable physical quantities are described by operators obeying a non-commutative law of multiplication. This is true, for example, of the coordinates and the components of the momentum of a mass point. It is true also of other types of observables like, for example, any component of the electric field at a given point of space. In quantum electrodynamics the components of the field or the potentials at a position in space are considered as operators which in general do not commute with each other. A field, however, is a system with an infinite number of degrees of freedom and from this fact many complications arise.

No attempt will be made here to give a complete description of the quantization procedure adopted; it is described in detail in more specialized publications. Only the simplest ideas of the radiation theory will be outlined. For additional explanations see Appendix 1.

Since the early attempts at setting up the statistics of radiation it has been customary to talk of radiation oscillators. The electromagnetic field enclosed in a cavity with perfectly reflecting walls is capable of oscillating according to a number of different modes with different characteristic frequencies. Each mode can be excited independently of the others and has properties quite similar to those of an oscillator. In particular, one of the modes can take up an amount of energy:

$$\hbar\omega_s(n_s + 1/2) \tag{1}$$

where ω_s is the angular frequency of the mode and $n_s = 0, 1, 2 \cdots$. The additional term $\hbar\omega_s/2$, the so-called zero-point energy, can be neglected as a non-essential additive constant to the energy.[1]

---

1. Actually, this constant is infinitely large. The number of oscillators of frequency between ω and $\omega + d\omega$ is given by (7). Consequently the total amount of zero-point energy is

With this renormalization the total energy content of the radiation field may be written

$$W = \sum_s \hbar\omega_s n_s. \tag{2}$$

Each term of the sum represents the contribution to the total energy of one vibrational mode.

In the language of the photons (2) indicates that there are n_s photons of energy $\hbar\omega_s$. Each one of them is thought of as a corpuscle with a momentum p_s related to the wave length λ_s by the de Broglie relationship:

$$p_s = \frac{2\pi\hbar}{\lambda_s} = \frac{\hbar}{\bar{\lambda}_s} = \frac{\hbar\omega_s}{c}. \tag{3}$$

The representation of the electromagnetic field in terms of oscillators is incomplete. It is suitable to represent radiation phenomena but does not include, for example, an electrostatic field. One can show, however, that the radiation phenomena can be treated separately from the electrostatic phenomena. A complete description of electrodynamics is obtained by considering on one hand the radiation field due to the superposition of transversal waves of all frequencies and on the other hand the Coulomb forces between electric charges. In the present discussion we shall be primarily interested in the behavior of the radiation field and therefore we shall limit ourselves to the transversal waves.

As long as no perturbation disturbs the electromagnetic field the quantum numbers n_s of the radiation oscillators will be constants and there will be no change in the number of photons. A perturbation will induce transitions whereby the number n_s may either increase (emission or creation of quanta) or decrease (absorption or destruction of quanta). In order to understand this fundamental point we shall discuss first an ordinary linear oscillator. This can be excited to the nth quantum state. The excitation energy of this state excluding the rest energy is $\hbar\omega n$. We say that the excitation amounts to n quanta of energy $\hbar\omega$ each. The number n of quanta will be a constant as long as the oscillator is left alone.

$$\int_0^\infty \frac{\hbar\omega}{2} \frac{\Omega\omega^2}{c^3\pi^2} \, d\omega.$$

This integral is obviously divergent at large frequencies. This is the first but not the worst example of infinities that one encounters in field theories.

Perturbations, however, may either increase or decrease the quantum number n. According to the general principles of quantum mechanics transitions will occur from an initial to a final value of the quantum number n when the matrix element of the perturbation corresponding to these values of the quantum number is different from zero. For example, if the perturbation is proportional to the abcissa x of the oscillator, the possible transitions will be those between values of n for which the matrix element of x does not vanish. These matrix elements are calculated in all elementary textbooks on quantum mechanics. They are different from zero only for transition of the quantum number from n to either $n + 1$ or $n - 1$. For a linear oscillator of mass m and frequency ω the only non-vanishing matrix elements of the coordinate x are

$$x(n \rightarrow n - 1) = \sqrt{\frac{\hbar}{2m\omega}} \sqrt{n} \,;$$

$$x(n \rightarrow n + 1) = \sqrt{\frac{\hbar}{2m\omega}} \sqrt{n + 1} \,. \tag{4}$$

The processes of creation and destruction of photons in the radiation theory are closely tied to this property of the oscillator. Indeed, the radiation field is equivalent to an assembly of linear oscillators. Transitions in which the excitation number n_s of one of the radiation oscillators increases are processes in which photons are created (emission of radiation). Transitions in which n_s decreases describe the destruction of photons (absorption of radiation).

In working out a quantitative radiation theory one finds it simpler to describe the field in terms of the vector potential A rather than of the electric and magnetic fields. As long as the discussion is limited to the radiation theory the scalar potential can always be assumed to be zero since an electromagnetic wave can be described by the vector potential only. The vector potential $A(r)$ at a point r is the sum $\Sigma A_s(r)$ of the vector potentials A_s contributed by the various modes.

A_s represents the field of the sth mode. Its magnitude is proportional to the coordinate of the radiation oscillator number s. Like this coordinate A_s has matrix elements inducing transitions from n_s to $n_s \pm 1$. The vector potential $A(r)$ at a given position in space is the sum of the quantities $A_s(r)$. $A(r)$ will therefore also be an operator having non-vanishing matrix elements for transitions in which the quantum number n_s of one of the oscillators changes to $n_s \pm 1$.[2]

Note that while $A(r)$ is an operator, the vector r that defines the position in space at which the vector potential is observed is an ordinary classical vector.

The actual values of the matrix elements can be obtained (see Appendix 1) by expressing A in terms of the oscillator coordinates and using (4). One finds the following result for the matrix elements of the observable $A(r)$ corresponding to transitions in which one photon $\hbar\omega_s$ is either created or destroyed:

$$A(r)(n_s \to n_s + 1) = \epsilon_s \frac{\sqrt{2\pi\hbar c}}{\sqrt{\Omega\hbar\omega_s}} e^{-(i/\hbar)p_s \cdot r} \sqrt{n_s + 1}$$

$$A(r)(n_s \to n_s - 1) = \epsilon_s \frac{\sqrt{2\pi\hbar c}}{\sqrt{\Omega\hbar\omega_s}} e^{(i/\hbar)p_s \cdot r} \sqrt{n_s}. \tag{5}$$

Here p_s is the momentum of the photon. Its magnitude is

$$|p_s| = \hbar\omega_s/c \tag{6}$$

while ϵ_s is a unit vector perpendicular to p_s and pointing in the direction of the polarization.

The formula giving the number of oscillation modes of frequency between ω and $\omega + d\omega$ will also be given:

$$dN = \frac{\Omega}{\pi^2 c^3} \omega^2 \, d\omega = 2\Omega \frac{4\pi p^2 \, dp}{8\pi^3 \hbar^3}. \tag{7}$$

In the last form of dN the factor $\Omega \times 4\pi p^2 \, dp$ is the volume element of phase space. This, divided by the cube of Planck's constant $h = 2\pi\hbar$, gives the number of modes except for the factor 2, due to the two possible polarization directions.

The second of the two expressions (5) contains the factor

$$u_s = \frac{1}{\sqrt{\Omega}} e^{(i/\hbar)p_s \cdot r}$$

which can be regarded as the eigenfunction of a photon of momentum p_s (normalized plane wave). This is a particular case of a general rule:

2. In the photon language a change from n_s to $n_s + 1$ means that a photon of the corresponding frequency has been created, and a change from n_s to $n_s - 1$ means that a photon has been destroyed.

The matrix element for the destruction of a particle is proportional to the eigenfunction of the state of the particle that is destroyed.

A similar rule is:

The matrix element for the creation of a particle is proportional to the complex conjugate of the eigenfunction of the state of the particle that is created.

This second rule is exemplified by the first formula (5) which gives the matrix element for the creation of a photon and is proportional to

$$u_s^* = \frac{1}{\sqrt{\Omega}}\, e^{-(i/\hbar)\, p_s \cdot r}.$$

3. SCALAR FIELD WITH MASS

The photons behave like particles of rest mass zero as is indicated by the relationship between energy and momentum

$$\hbar\omega_s = cp_s \tag{8}$$

that holds for them. This relationship is a consequence of the D'Alembert equation

$$\nabla^2 A - \frac{1}{c^2}\frac{\partial^2 A}{\partial t^2} = 0 \tag{9}$$

from which it follows that the photons travel with the velocity of light.

For this reason the field equation (9) cannot be used for the pion field. The pions have a rest mass different from zero and travel with a velocity less than that of light. Like the photons they are believed to obey the Bose-Einstein statistics since according to the Yukawa theory pions can be emitted or absorbed by nucleons. In this section and in the next a type of field will be discussed whose quanta have a non-vanishing rest mass. This field might be adopted for the description of pions. The case of neutral pions will be discussed in this section and that of charged pions in the next. In both cases these particles will be assumed to have zero spin so that they may be represented by a scalar field.

The simplest field equation for the scalar pions is the Klein-Gordon equation

$$\nabla^2 \varphi - \frac{1}{c^2} \frac{\partial^2 \varphi}{\partial t^2} - k^2 \varphi = 0 \tag{10}$$

which differs only by the last term from the D'Alembert equation. The field amplitude φ is a scalar.[3]

Equation (10) has plane-wave solutions of the form

$$a \exp i(x/\lambda - \omega t).$$

Substituting in (10) one obtains

$$-\frac{1}{\lambda^2} + \frac{\omega^2}{c^2} - k^2 = 0. \tag{11}$$

As in the case of the electromagnetic field, the vibrations of the field φ can be described as a superposition of fundamental modes with different characteristic frequencies ω_s. Each mode behaves like a linear oscillator of frequency ω_s. Omitting again the zero-point energy, the energy levels of this oscillator will be $\hbar\omega_s n_s$. Also in this case the integral number n_s is interpreted as the number of quanta in the mode s, each having the energy $w_s = \hbar\omega_s$. The momentum p_s of each quantum is given by the de Broglie relationship $p_s = \hbar/\lambda_s$. Using this value of p_s and (11) the following relationship is obtained between the energy and momentum of a quantum:

$$w_s = \hbar\omega_s = \sqrt{c^2 p_s^2 + \hbar^2 c^2 k^2}. \tag{12}$$

This is equivalent to the relativistic energy-momentum relationship

$$w = \sqrt{c^2 p^2 + m^2 c^4} \tag{13}$$

for a particle with rest mass

$$m = \hbar k/c. \tag{14}$$

3. Scalar quantities are usually classed as scalars proper and pseudo-scalars. When the space coordinates are reflected with respect to the origin a scalar remains unchanged and a pseudoscalar changes its sign. According to the assumptions made on the behavior of φ one can construct either a scalar or a pseudoscalar theory of the pions. There is no difference between these two theories except in the allowable forms of interaction of the pions with the nucleons. (See Section 8.)

Most of the discussion of Section 2 and Appendix 1 can be repeated for the scalar field φ. In particular the amplitude $\varphi(r)$ of the field at a position r is an operator with matrix elements corresponding to the creation or to the destruction of a quantum. The matrix elements of $\varphi(r)$ are

$$\varphi(r)(n_s \to n_s - 1) = \frac{\hbar c}{\sqrt{2\Omega w_s}}\, e^{-(i/\hbar)p_s \cdot r}\, \sqrt{n_s} \tag{15}$$

$$\varphi(r)(n_s \to n_s + 1) = \frac{\hbar c}{\sqrt{2\Omega w_s}}\, e^{(i/\hbar)p_s \cdot r}\, \sqrt{n_s + 1}. \tag{16}$$

The matrix elements (15) and (16) are quite analogous to (5). Since φ is a scalar, of course we have no polarization vector ϵ. The difference by the factor $\sqrt{4\pi}$ is due to having used Heavyside units for φ and non-rationalized units for the electromagnetic field. When the field is included inside a volume Ω only a discrete series of momentum values is allowable. The formula giving the number of allowable momentum values between p and $p + dp$ is

$$dN = \frac{\Omega \times 4\pi p^2\, dp}{(2\pi\hbar)^3} = \frac{\Omega p^2\, dp}{2\pi^2 \hbar^3}. \tag{17}$$

This formula commonly used in statistical mechanics is quite similar to (7) and differs from it only by a factor 2, due to the two different polarization possibilities of the photons.

4. FIELD OF CHARGED SCALAR PARTICLES

As a rule, particles without electric charge are quanta of a field with real components. For example, the photons are the quanta of the electromagnetic field. Instead, it is found that a field whose quanta are electrically charged has complex amplitude. This is true also for particles which, though neutral, have some electromagnetic property—for example, the neutron with its magnetic moment. This property is quite general and is related to the requirement of gauge invariance.[4]

For this reason the real field φ of the previous section has neutral particles as its photons and therefore could be used as the field of the neutral pions. Pauli and Weisskopf have shown, however, that a scalar complex quantity φ obeying the Klein-Gordon equa-

4. See for example Pauli, *Rev. of Mod. Phys.*, *13*, 203, 1941.

tion may be used to describe a field whose quanta are electrically charged. A complex field of this type will be used for the charged pions. The quanta of this complex field can be charged both positively and negatively. For each allowable value p_s of the momentum there may be n_s^+ positive and n_s^- negative particles. Omitting again the zero-point energies, the total energy in the field will be

$$W = \sum_s w_s(n_s^+ + n_s^-). \tag{18}$$

Both the complex quantity φ and its complex conjugate φ^* will be important. The quantities $\varphi(r)$ and $\varphi^*(r)$ are operators whose matrix elements connect states in which the number of particles changes by ± 1. More precisely, Pauli and Weisskopf have found that $\varphi(r)$ induces transitions in which either n_s^+ decreases by one unit or n_s^- increases by one unit. The opposite holds for the conjugate field $\varphi^*(r)$ which produces either the transitions $n_s^+ \to n_s^+ + 1$ or $n_s^- \to n_s^- - 1$. In other words, φ is the operator that reduces the total charge by one unit, either by destroying a positive particle or by creating a negative one. φ^* has the opposite effect.

Since as far as we know the electric charge is always conserved, we expect that neither φ nor φ^* will ever appear alone in any term of the Hamiltonian. They will always be associated with other quantities so that the charge will be conserved. For example, φ and φ^* alone would be inacceptable. But their product $\varphi^*(r)\varphi(r)$ is acceptable.

Except for these differences, the matrix elements of φ and φ^* are quite similar to (15) and (16). They are

$$\varphi(r)(n_s^\pm \to n_s^\pm \mp 1) = \frac{\hbar c}{\sqrt{2\Omega w_s}} e^{\pm(i/\hbar)p_s \cdot r} \begin{cases} \sqrt{n_s^+} \\ \text{or} \\ \sqrt{n_s^- + 1} \end{cases}$$

$$\varphi^*(r)(n_s^\pm \to n_s^\pm \pm 1) = \frac{\hbar c}{\sqrt{2\Omega w_s}} e^{\mp(i/\hbar)p_s \cdot r} \begin{cases} \sqrt{n_s^+ + 1} \\ \text{or} \\ \sqrt{n_s^-} \end{cases} \tag{19}$$

The formula giving the number of allowable values of the momentum p_s is identical to (17).

5. PARTICLES OBEYING THE
PAULI PRINCIPLE

The quanta of the electromagnetic field and those of the scalar field, both neutral and charged, obey the Bose-Einstein statistics. This is evident because any integral number n_s of particles can be found in a state of given momentum p_s.

A different type of theory is needed for the many elementary particles that are known or believed to obey the Pauli principle. They are electrons and positrons, protons, neutrons, neutrinos, and probably muons. Particles obeying the Pauli exclusion principle also can be described in terms of a field theory by the so-called second quantization procedure of Jordan and Wigner, for which the number of particles in a quantum state can be only 0 or 1.

First, relativity corrections and spin will be neglected. One finds in this case that the role of field quantity is played by the probability amplitude ψ obeying the time-dependent Schroedinger equation which in the absence of external forces is

$$\frac{\partial \psi}{\partial t} = \frac{i\hbar}{2m} \nabla^2 \psi. \tag{20}$$

In the ordinary Schroedinger theory this equation is construed to describe the state of a single particle (electron). The amplitude of probability for observation of the electron at position r is then $\psi(r)$. In the field theory of the electrons equation (20) is still used, formally unchanged but with a very different meaning. In the case of the Klein-Gordon equation we have treated the field φ as an operator which has matrix elements given by (15) and (16) connecting states in which the number of particles changes by ± 1. In a similar way the field quantity ψ also will be considered as an operator. For electrons obeying the Pauli principle, however, the number of particles in a state of given momentum p_s can be only 0 or 1 as long as the spin is neglected. In other words, the occupation numbers n_s can be only 0 or 1. The operator properties of the electron field must be such as to yield this property (see Appendix 2). As in the case of the scalar field φ one finds that both $\psi(r)$ and $\psi^*(r)$ are operators which have matrix elements connecting states for which the occupation number n_s changes by one unit, either from 1 to 0 or from 0 to 1. One finds (see Appendix 2) the following matrix elements:

$$\psi(r)(1 \rightarrow 0) = \pm \frac{1}{\sqrt{\Omega}} e^{(i/\hbar)p_s \cdot r}$$

$$\psi^*(r)(0 \rightarrow 1) = \pm \frac{1}{\sqrt{\Omega}} e^{-(i/\hbar)p_s \cdot r} \tag{21}$$

The $+$ or $-$ sign should be adopted according to a somewhat complicated rule,[5] which need not concern us here since in most cases the square modulus of the matrix element will be used.

It should be noted that ψ acts as a destruction operator (transition $1 \rightarrow 0$) and ψ^* as a creation operator (transition $0 \rightarrow 1$).

When no forces act on the particles their representation in terms of states of given momentum (plane waves) is the most convenient. When the spin properties are neglected this leads to the matrix elements (21). When external forces act on the particles, however, it is preferable to analyze in terms of eigenfunctions of the particles in the field of the external forces rather than in plane waves.

Instead of characterizing a state by indicating the number n_s of electrons of a given momentum p_s, we characterize the state by indicating the number m_i of electrons in the quantum state (i) of the electron in the external field. Again the Pauli principle will limit the values of the occupation numbers m_i to only 0 or 1. The matrix element of the field $\psi(r)$ and of the complex conjugate quantity $\psi^*(r)$ will connect states in which m_i changes by ± 1 as follows (see Appendix 2):

$$\psi(r)(1 \rightarrow 0) = \pm u_i(r)$$

$$\psi^*(r)(0 \rightarrow 1) = \pm u_i^*(r) \tag{22}$$

where $u_i(r)$ is the eigenfunction of the ith state of an electron in the given force field. Notice that the matrix elements (21) are a particular case of (22) when no forces act on the electron because the expressions on the right-hand sides of (21) are the plane waves

5. One must order the states according to some arbitrary rule, for example, increasing values of p_s. Of the states that precede s a certain number will be occupied. The sign to be chosen is $+$ or $-$ according to whether this number is even or odd.

that represent the normalized eigenfunctions of an electron on which no forces act.

In most calculations of orders-of-magnitude formulas (21) or (22) are used. However, the case in which the relativistic Dirac equation is adopted will also be briefly outlined. In this case instead of (20) the field equation will be the Dirac equation

$$\frac{1}{c}\frac{\partial \psi}{\partial t} = \alpha \cdot \nabla \psi + \frac{imc}{\hbar}\beta \psi \tag{23}$$

where α_x, α_y, α_z, β are the well-known Dirac matrices. In this case the field ψ is no longer a scalar but is a four-component spinor.

For each momentum p_s there are four states corresponding to the two spin orientations and to positive or negative energy:

$$w_s = \pm\sqrt{c^2 p_s^2 + m^2 c^4}. \tag{24}$$

The four-component wave functions of one of these states can be written in the form of a plane-wave solution of the Dirac equation:

$$\frac{1}{\sqrt{\Omega}} B_\lambda^{(\mu)} e^{(i/\hbar)p_s \cdot r}$$

where the four-valued index μ specifies the spin and the sign of the energy of the state, and the four-valued index λ denotes the four components of the wave function. The quantities B are dimensionless and are of the order of magnitude 1.[6]

In the electron-field theory all four components $\psi_\lambda(r)$ and their complex conjugates $\psi_\lambda^*(r)$ are operators. They induce destruction and creation transitions of the occupation number n_s from 1 to 0 or from 0 to 1 with matrix elements

$$\psi_\lambda(r)(1 \to 0) = \pm\frac{1}{\sqrt{\Omega}} B_\lambda^{(\mu)} e^{(i/\hbar)p_s \cdot r}$$

$$\psi_\lambda^*(r)(0 \to 1) = \pm\frac{1}{\sqrt{\Omega}} B_\lambda^{*(\mu)} e^{-(i/\hbar)p_s \cdot r}. \tag{25}$$

6. When the electron moves with velocity small compared with c one finds that two of the four quantities B are of the order of 1 and two are of the order of magnitude v/c. For a positive energy state the components 3 and 4 are large and the components 1 and 2 are small, and for a state of negative energy the opposite is the case.

In most of the applications of this formula we shall be concerned merely with orders of magnitude and shall therefore replace B with its order of magnitude 1. The matrix elements then become identical to (21).

In the Dirac hole theory all the positive-energy states in the vacuum are unoccupied and all the negative-energy states are fully occupied by one electron per state. Creation of a positron corresponds to the destruction of an electron in one of these negative-energy states. Consequently the operator ψ serves the dual purpose either of destroying an electron or of creating a positron. The operator ψ^* has the opposite effect.

CHAPTER TWO

Interaction of the Fields

6. GENERAL TYPES OF INTERACTION

The fields discussed in Chapter I will be used in the applications to follow for the interpretation of a number of properties of the following elementary particles:

The photon (symbol γ) is described by the radiation field of Section 2 with amplitude A the vector potential.

The pions, charged and neutral (indicated by symbols Π^+, Π^-, Π^0), will be assumed for simplicity to have spin zero. The charged pions are described by a complex scalar field as in Section 4 and the neutral pions by a real scalar field as in Section 3. As a rule, the letter indicating the particles will be used to indicate the field amplitude as well. For pions, for example, the letter Π will denote the amplitude of the field in place of the letter φ used in Sections 3 and 4.

Electrons (symbol e), protons (symbol P), neutrons (symbol N), muons (symbol μ), neutrinos (symbol ν) are all assumed to obey the Dirac field equation of Section 5. In estimates of order of magnitude the spinor character will be disregarded and the simplified expressions (21) for the matrix elements of the field amplitudes will be applied. The symbols of the particles will be used to represent either the particle or the amplitude of the corresponding field.

Interactions between various particles are responsible for a variety of phenomena such as the scattering of two colliding elementary particles and the more complex events in which some particles disappear and some are created. Two particles may

16

interact only when they occupy the same position in space (contact interaction) or when they are separated by a finite distance. This last case is usually interpreted not as an actual action at a distance but as the effect of a field that transmits the force from one to the other particle. In the quantum interpretation this field will have its own quanta and the interaction between the two original particles will be described as a process in which a quantum of this field is emitted by one of the particles and reabsorbed by the other. This gives rise to a mutual energy of the two particles that is a function of their distance and represents the potential of the force field. The trend is to consider more satisfactory a theory with only contact interactions which is more easily brought into a relativistically invariant form.

In almost all field theories that are at present under discussion the interaction energy of two or more types of particles is represented as the volume integral of an interaction energy density. This has usually the form of a product of amplitudes of the various fields.

In the case of the interaction of electrons and the radiation field the general form of the density of interaction energy is suggested by its expression in classical electrodynamics. This is given by the scalar product $A \cdot J$ of the vector potential A and the current density J of the electrons. Since the scalar potential vanishes in a radiation field this is the only interaction term. Otherwise the product of electric density and scalar potential also should be added. The current density J can be expressed in terms of the amplitude ψ of the electron field. Neglecting spin and relativity, one can express J as in the elementary Schroedinger theory by

$$J = \frac{e\hbar}{2imc} (\psi^* \nabla \psi - (\nabla \psi^*)\psi). \tag{26}$$

The interaction energy is thus written in the form

$$\mathcal{K} = \frac{e\hbar}{2imc} \int (\psi^* \nabla \psi - (\nabla \psi^*)\psi) \cdot A \, d\Omega. \tag{27}$$

In order to calculate the transitions due to the interaction (27) of electrons and photons, its matrix elements will be needed. In computing them the first term in (27) will be considered first. It contains the operator A which can cause the destruction or creation of a photon according to (5), and the operators ψ and ψ^*

which, according to (21), produce respectively the destruction and the creation of an electron. Let p_s be the momentum of the photon that is emitted (or absorbed), p_1 be the momentum of the electron that is destroyed, and p_2 the momentum of the electron that is created. Destroying an electron of momentum p_1 and creating another one of momentum p_2 is of course equivalent to a transition in which an electron changes its momentum from p_1 to p_2 . Therefore the matrix elements of (27) will connect the initial state to a final state that differs from it because a photon of momentum p_s either has been created or destroyed and at the same time an electron has changed its momentum from p_1 to p_2. We consider for example the case in which the initial state contains no photon of momentum p_s : in other words, the initial value of the occupation number n_s of the photons in the given state is zero. If a photon of momentum p_s is created the occupation number of the final state will be $n_s = 1$. According to the first formula (5) the corresponding matrix element of the vector potential is

$$\epsilon_s \, \frac{\sqrt{2\pi}\;\hbar c}{\sqrt{\Omega \hbar \omega_s}} \, e^{-(i/\hbar)\,p_s \cdot r}. \tag{28}$$

According to (21) the operator ψ that causes the destruction of an electron of momentum p_1 will contribute to the matrix element the factor $\pm e^{(i/\hbar)p_1 \cdot r}/\sqrt{\Omega}$. According to (27) we need the gradient of this factor. This gives merely an additional factor $ip_1/\hbar$. Finally, the operator ψ^* that is responsible for the creation of an electron of momentum p_2 contributes in (27) the factor $\pm e^{-(i/\hbar)p_2 \cdot r}/\sqrt{\Omega}$.

In conclusion the integrand of the first term of (27) contributes to the matrix element the term[1]

$$\frac{e\hbar}{2imc} \frac{1}{\sqrt{\Omega}} \, e^{-(i/\hbar)p_2 \cdot r} \frac{1}{\sqrt{\Omega}} \, e^{(i/\hbar)p_1 \cdot r} \frac{i}{\hbar} \, (p_1 \cdot \epsilon_s) \frac{\sqrt{2\pi}\;\hbar c}{\sqrt{\Omega \hbar \omega_s}} \times$$
$$\times \, e^{-(i/\hbar)p_s \cdot r} = \frac{e}{2m\Omega} \sqrt{\frac{2\pi\hbar}{\Omega \omega_s}} \, (p_1 \cdot \epsilon_s) e^{(i/\hbar)(p_1 - p_2 - p_s) \cdot r}. \tag{29}$$

This expression must be integrated over the normalization volume Ω. The integration is immediate, since the only space-dependent factor in the above expression is the exponential. Its integral

---

1. The sign of this expression is correctly given in (29) in the practically important case that one electron only is involved.

over the large volume Ω will be different from zero only when the vector

$$\vec{p_1} - \vec{p_2} - \vec{p_s} = 0 \tag{30}$$

vanishes, because otherwise the integrand averages to zero. The condition (30) expresses the conservation of momentum: The matrix element vanishes unless the initial momentum p_1 of the electron is equal to the vector sum of the momenta p_s of the photon and p_2 of the electron in the final state. If the momentum is conserved the exponential is equal to 1 and its volume integral is Ω. The total contribution to the matrix element of the first term of (27) is then

$$\frac{e}{2m} \sqrt{\frac{2\pi\hbar}{\Omega\omega_s}} \, (p_1 \cdot \epsilon_s). \tag{31}$$

The second term of (27) yields a similar contribution with p_1 changed into p_2. Since (27) is a non-relativistic formula, the expressions p_1/m and p_2/m are the velocities v_1 and v_2 of the electron before and after the transition. The matrix element is therefore

$$\frac{e}{2} \sqrt{\frac{2\pi\hbar}{\Omega\omega_s}} \, (\vec{v_1} + \vec{v_2}) \cdot \epsilon_s. \tag{32}$$

It can be proved in a similar way that the matrix element corresponding to the destruction of a photon also is given by (32), provided the momentum is conserved between initial and final states. Otherwise this matrix element also vanishes.

The two processes which have been considered correspond to the emission or the absorption of light accompanied by a change of momentum of the electron in comformity with the momentum conservation. It should be noticed, however, that the fact that these processes have a non-vanishing matrix element does not mean that the processes actually happen. It is well known that a free electron cannot emit radiation because in this case it is impossible to satisfy at the same time the conservation of momentum and of energy. In Section 10 the methods for calculating the probabilities of transition and in particular the role of the conservation of energy will be discussed. (See also Appendix 4.)

Formula (32) gives the correct order of magnitude of the matrix element also in the relativistic case in which the electrons obey the

Dirac equation. The expressions for the current density and the interaction energy are then

$$J = -e\tilde{\psi}\alpha\psi \tag{33}$$

$$\mathcal{K} = \int J \cdot A \, d\Omega = -e \int \tilde{\psi}\alpha\psi \cdot A \, d\Omega \tag{34}$$

where $\tilde{\psi}$ is the transposed conjugate of the four-component Dirac function ψ.

In the relativistic case both the Dirac matrices α and the quantities B of (25) have the order of magnitude 1. Using (5) one finds that the matrix element for the creation or the destruction of a single photon is of the order of magnitude

$$ec \sqrt{\frac{2\pi\hbar}{\Omega\omega_s}}. \tag{35}$$

When the velocities are close to c this expression differs from (32) only by a factor of the order of unity.

For particles obeying the Bose-Einstein statistics, like the pions for example, the expression of the current density and consequently also the form of the electromagnetic interaction are somewhat different from (27) or (34). One finds, however, that in the simple processes which we are going to discuss the order of magnitude of the matrix elements for processes of creation and destruction of a photon is given also by (32).

7. CONSERVATION OF MOMENTUM

It has been seen that the conservation of momentum in the case of the electromagnetic interaction is a necessary condition in order to have a non-vanishing matrix element. The same is true also for the other field interactions that we shall encounter. It will be seen that all interactions are volume integrals of terms containing factors equal to the amplitudes of the various fields (or sometimes of their derivatives). It is seen from (5), (15), (19), (21), (25) that for all types of fields the space dependence of the matrix elements of the various field components is contained in an exponential factor $exp\,(\pm ip \cdot r/\hbar)$ where p is the momentum of the particle that is created or destroyed. The sign is always $+$ for the destruction matrix elements and $-$ for the creation elements. The volume integral of the energy density involves therefore merely the integral over the space Ω of a product of exponentials which may be written

$$\int e^{(i/\hbar)(\pm p_1 \pm p_2 \pm \ldots \pm p_n) \cdot r}. \tag{36}$$

This integral is different from zero only when the vector

$$\pm \vec{p}_1 \pm \vec{p}_2 \pm \cdots \pm \vec{p}_n = 0 \tag{37}$$

because otherwise the integrand averages to zero. When this condition is fulfilled the integrand is 1 and the integral becomes equal to the normalization volume Ω. The above equation expresses the momentum conservation because in it the momenta of the particles destroyed appear with the plus sign and those of the particles created appear with the minus sign. Hence the vector sum of the momenta of the particles created equals the vector sum of the momenta of the particles destroyed. One obtains therefore the following theorem and the following rule:

Theorem of the momentum conservation. Two states that are connected by a non-vanishing matrix element of the interaction energy operator have equal momenta.

Rule for calculating volume integrals. In calculating the volume integral of the energy density the momentum conservation is required. The integral is then obtained by substituting the volume Ω for the volume integral of the product of all the exponential factors. These properties greatly simplify the computation of the matrix elements.

For further discussion of the energy-momentum conservation in field theories see Appendix 4.

8. YUKAWA INTERACTIONS

In this section and the next the main interactions between elementary particles will be enumerated. We begin with the Yukawa interaction between pions and nucleons. The assumption is made that there are matrix elements generating the following transformations:

$$\text{(a) } P \rightleftarrows N + \Pi^+ \qquad \text{(b) } N \rightleftarrows P + \Pi^-$$
$$\text{(c) } P \rightleftarrows P + \Pi^0 \qquad \text{(d) } N \rightleftarrows N + \Pi^0. \tag{38}$$

In the original Yukawa theory only the reactions involving charged pions were considered. Recent evidence makes it almost certain that a neutral pion also exists, and accordingly the two reactions (38c) and (38d) have been added. If the reactions (38) are correct one would expect the pion to obey Bose statistics and have integral spin. If the spin is 0 a representation with a scalar field as discussed in Sections 3 and 4 is plausible. If the spin were 1 a vector field should be adopted instead. In what follows a spin 0 will be assumed.

This still leaves open the possibilities that the field amplitude, which will be denoted by Π, may be a scalar or a pseudoscalar. It is fashionable to assume the latter since one finds that it leads to nuclear forces less in disagreement with experiment. We shall follow this fashion.

When Π represents a neutral pion field it will be a real quantity such as in Section 3. Positive and negative pions are represented instead by a complex field. One should of course use a different notation for the two cases. In order to keep our formulas simpler this will not be done. The point will be clarified whenever necessary. The discussion that follows will refer to the example of charged pions for which Π is a complex field.

From (19) and (21) it follows that the reactions (38a and b) can be produced by terms in the interaction energy containing the factors Π^*N^*P and ΠP^*N. In computations of orders of magnitude the interaction energy density will be taken proportional to these factors.

For this particular case, however, it may be instructive to look somewhat more closely into the actual form of the possible interaction term.[2] Since Π and Π^* have been assumed to be pseudoscalars, the simplest invariant combinations of the desired type are obtained by associating the pion amplitudes with a pseudoscalar expression patterned according to (3) in Appendix 4. The simplest

2. The reader interested in this part of the discussion should first read Appendix 4. Otherwise go over to formula (41).

possibility for the interaction term in the Lagrangian density is therefore the Hermitian operator

$$-e_2'(\Pi \tilde{P}\beta\alpha_1\alpha_2\alpha_3 N + \Pi^*\tilde{N}\beta\alpha_1\alpha_2\alpha_3 P).$$

Here e_2' is a coefficient that determines the strength of the inter-action and has the dimensions of an electric charge. It is found that the above term in the Lagrangian leads to a Hamiltonian density equal to it except for the sign. We can therefore write the Hamiltonian

$$\mathcal{H}' = e_2' \int (\Pi \tilde{P}\beta\alpha_1\alpha_2\alpha_3 N + \Pi^*\tilde{N}\beta\alpha_1\alpha_2\alpha_3 P) \, d\Omega. \qquad (39)$$

This is not the only possibility, because if Π is a pseudoscalar its gradient is a pseudovector. Therefore an invariant expression can be formed as the four-dimensional vector product of the gradient of Π or Π^* and a pseudovector formed with P and N according to the pattern of (5) in Appendix 4. This yields in the Hamiltonian a term

$$\mathcal{H}'' = \frac{\hbar e_2''}{\mu c} \int \left(\nabla\Pi \cdot \tilde{P} \, \frac{\alpha \times \alpha}{i} \, N + \text{compl. conj.} \right) d\Omega \qquad (40)$$

where μ is the pion mass. The factor $\hbar/\mu c$ has been written explicitly in the coefficient so that the constant e_2'' again will have the dimensions of an electric charge.

In computing orders of magnitude simplified expressions will be used. Instead of using (39) or (40) or a combination of them one can simplify the calculations by neglecting the spin properties of the nucleons and by writing the interaction term in the form

$$\mathcal{H} = e_2 \int (\Pi P^*N + \Pi^*N^*P) \, d\Omega \qquad (41)$$

where P, P^*, N, N^* have matrix elements computed according to (21).

The errors introduced by adopting (41) are various. First of all, the results are not relativistically invariant. Therefore one cannot expect the behavior of particles with energy very large compared to their rest energy to be represented properly. Also at low velocities errors may be expected. Although the Dirac matrices and their products have elements of the order of unity, factors of the order v/c frequently appear in their expectation values. The same is true of the quantity $(\hbar/\mu c)\nabla$ appearing for example in (40).

The main features that are lost in substituting (41) for (39) or (40) are the qualitative understanding of the spin dependence of the nuclear forces and of the contribution of the pion field to the magnetic moment of the nucleons. These effects of course would not be obtainable from a spinless expression like (41).

These disadvantages are in part compensated by the great simplicity of the calculations based on (41). Also one should observe that a convincing form of the interactions is not known, except in the case of the interaction between electrons and electromagnetic field. Indeed, no experimental evidence can be quoted to prove that interactions like (39) or (40) are correct. This unsatisfactory situation is probably due in part to the fact that no proper mathematics has been evolved capable of handling interaction terms unambiguously. The procedure most used currently is to regard the interaction terms as perturbations and to compute the probabilities of transition to the lowest approximation that yields the desired result. Even within this limited scope one frequently encounters divergent expressions that are estimated by cut-off procedures of dubious reliability.

Under these conditions one wonders whether it pays to spend time and effort in computing some terms exactly while other divergent terms are neglected. The fact that field theories often give results in agreement with experiment at least as to the order of magnitude makes it appear likely that the final theory will bear some similarity to the present attempts. Perhaps the introduction of a finite size of the elementary particles or even a granular geometry such as is suggested by Heisenberg and Snyder may be clues to the solution.

These somewhat disparaging remarks as to the status of present-day field theories do not apply to quantum electrodynamics. This discipline has progressed to a point where it gives us a detailed understanding of the photon-electron interaction. Great strides have been made in it in the last few years. Since, however, there exist excellent monographs on quantum electrodynamics, we will not give it much attention here but refer the reader to more specialized publications.

The Yukawa processes (38) have a very important consequence.

According to (38b) for example, a neutron is convertible into a proton and a negative pion. From the existence of a strong interaction between these two possible states it follows, as will be discussed in more detail in Section 11, that there is a continuous interchange between these two forms of the neutron. The neutron spends part of its time as neutron proper and part as a proton with a negative pion nearby. Similarly, part of the time a proton is a proton proper and part of the time it is a neutron with a positive pion in the vicinity. The anomalies of the magnetic moment of the nucleons already mentioned are interpreted as being due to the magnetic effects of the pions that are found a part of the time in the vicinity of the nucleon. Unfortunately the attempts to explain the magnetic anomalies quantitatively in this manner have failed, and nothing more than an agreement in the orders of magnitude has been achieved.

Formula (41) applies to the case of charged pions. The assumption is usually made that the interaction of the neutral pions with the nuclei is similar to it. In this case, however, Π is a real quantity and therefore Π and Π^* are the same. The present evidence seems to indicate that the coupling constants e_2 have approximately equal values for the interactions of the charged and the neutral pions. In this book the two interaction constants will be taken as equal.

We conclude this section by computing the matrix elements of the interaction (41) according to the rules at the end of Section 7. These formulas will later be used in the applications. The matrix elements for all processes (38) in which a pion is created or destroyed are obtained from the formulas (19) for calculating the matrix elements of Π and (21) for calculating the matrix elements of P and N. The integral over the exponential factor is replaced by Ω as explained in Section 7. The result is

$$e_2 \frac{\hbar c}{\sqrt{2\Omega w_s}} \frac{1}{\sqrt{\Omega}} \frac{1}{\sqrt{\Omega}} \Omega = \frac{e_2 \hbar c}{\sqrt{2\Omega w_s}} \tag{42}$$

where w_s is the total energy of the pion that has been either created or destroyed. In computing these formulas it has been assumed that the transition of the occupation number of the pion was either from 0 to 1 in a creation process or from 1 to 0 in a destruction process. Otherwise the factors $\sqrt{n_s + 1}$ or $\sqrt{n_s}$ appearing in (19) should be included.

9. OTHER INTERACTIONS

It is known experimentally that the pion is an unstable particle and that it decays spontaneously with a lifetime of about 2×10^{-8} seconds into a muon. Powell, who first observed this decay, found that the muon decaying from a pion at rest always has a constant range of about 600 microns in the photographic emulsion. From considerations of energy-momentum conservation one concludes that in the pion decay a neutral particle must also be emitted which leaves no photographic trace. It is consistent with the rather wide limits of experimental error to assume that this neutral particle has mass zero. The current assumption that we shall follow is that it is a neutrino. It is, however, more convenient to call this particle an anti-neutrino (symbol $\bar{\nu}$), which of course makes little difference. Therefore we shall postulate the reactions

$$\Pi^+ \rightleftarrows \mu^+ + \bar{\nu} \quad \text{and} \quad \Pi^- \rightleftarrows \mu^- + \nu. \tag{43}$$

An interaction term suitable to cause the first of these transitions should include the operator Π, which according to (19) destroys a positive pion, and the operators μ^* and ν, which according to (21) create a positive muon and destroy a neutrino, i.e., create an anti-neutrino. The same combination of operators will also cause the inverse of the second reaction (43) because the operator Π can create a negative pion. The operator μ^* may create a positive muon of the negative Dirac sea of these particles, that is, destroy a negative muon. Similarly the operator ν may destroy a neutrino. Since the interaction energy is a real quantity (Hermitian operator) its complex conjugate $\Pi^*\nu^*\mu$ should be added to the previous term. This complex conjugate is suitable to produce the second reaction (43) and the inverse of the first. The simplest form of interaction energy suitable to produce the reactions (43) and their inverses would therefore be

$$e_3 \int (\Pi\mu^*\nu + \Pi^*\nu^*\mu) \, d\Omega \tag{44}$$

where e_3 is a suitable interaction constant having the dimensions of an electric charge. The interaction term (44) is not relativistically nvariant and should be replaced by more elaborate expressions somewhat on the lines followed in discussing various forms of the Yukawa interaction. Since our aim is merely to calculate orders of magnitude we shall forego these complications in the interest of simplicity.

The matrix elements of the interaction (44) can be calculated with the usual procedure by combining the matrix elements of I, μ, ν obtained respectively from (19) and (21) and applying the rule of Section 7. One finds that the matrix elements for all processes (43) have equal values,[3] namely:

$$\frac{e_3 \hbar c}{\sqrt{2 \Omega w_s}} \tag{45}$$

where w_s is the energy of the pion. This formula will be used in Section 13 when the lifetime of the pion is discussed.

The interaction constant between electrons and the radiation field is the electronic charge e. The interaction constants e_2 of the Yukawa theory (formula (41)) and the constant e_3 for the process (43) all have the dimensions of an electric charge. Their actual values, however, are quite different, as will be seen in the next chapter.

We shall proceed now to discuss three other elementary processes which have interaction constants with the dimensions ergs $\times$ cm^3. These constants will be denoted by g_1, g_2 and g_3.

The first of these interactions is responsible for the beta-ray emission. This process has long since been interpreted as due to the elementary-particle reaction

$$N \rightleftarrows P + e + \bar{\nu}. \tag{46}$$

The formulas become more symmetric by assuming that an anti-neutrino and not a neutrino is associated with the emission of an electron, and the reaction (46) has been written accordingly. A transition of this type might be due to an interaction term in the Hamiltonian of the simplified form

$$g_1 \int (P^*Ne^*\nu + N^*P\nu^*e) \, d\Omega. \tag{47}$$

3. As in the case of the similar formula (42) the assumption has been made that the occupation numbers of the pions change either from 0 to 1 or from 1 to 0. Otherwise the factor $\sqrt{n+1}$ or $\sqrt{n}$ would appear in (45).

In this case all four particles are of the Pauli type and (21) wi apply. With the rules of Section 7 the matrix element correspondin to the interaction (47) is immediately found to be

$$g_1 \left(\frac{1}{\sqrt{\Omega}} \right)^4 \Omega = \frac{g_1}{\Omega}. \tag{48}$$

This formula, however, corresponds to a transition in which a four particles of reaction (46) are in states with definite momentu values (plane waves). In the applications also the formulas co responding to the case in which the states of the neutron, proton and electron are not representable as plane waves are frequentl used. These formulas could be obtained by using the expression (22) instead of (21) for the matrix elements of the field component The corresponding formulas, as well as some of the many attempt that have been made in order to determine the precise form of th beta-ray interactions in more detail than is given by the simplifie type (47), will be mentioned in Section 14.

We proceed to discuss another elementary interaction which responsible for the spontaneous decay of the muon. The muon an unstable particle. It has a mean life of 2.15×10^{-6} second and emits an electron. The measurements of Steinberger an Anderson indicate that the energy of the electron emitted is n always the same. This fact has been interpreted by Steinberge and Wheeler as an indication that the muon decays into thre particles, an electron and two neutral particles, usually assume to be neutrinos. On this assumption the reaction would be

$$\mu \rightleftarrows e + \nu + \bar{\nu}. \tag{49}$$

This reaction has been written for the case that one of the tw neutrinos is an anti-neutrino, a matter of little practical importanc A simplified interaction term leading to (49) could be

$$g_2 \int (\mu^* e \nu^* \nu + e^* \mu \nu^* \nu) \, d\Omega. \tag{50}$$

Also in this case the four particles are all of the Pauli type and (2 will be used for calculating the matrix elements of the field com ponents. One finds as in the case of the beta interaction that th matrix elements for the processes (49) are

$$\frac{g_2}{\Omega}. \tag{51}$$

The spontaneous decay (49) is not the only process by which a muon may vanish. There is evidence that a negative muon captured near a nucleus may disappear without emitting an electron. A reaction of the type

$$P + \mu^- \rightarrow N + \nu \tag{52}$$

is consistent with the known facts. As soon as the muon is captured near a nucleus it would react with one of the nuclear protons and the two particles would change into a neutron and a neutrino. One feature of this hypothesis is that from it follows that the nucleus in which the reaction takes place would not be strongly excited since most of the mass energy of the muon would be lost into neutrino energy. This would explain why the process as a rule does not lead to star formation.

In the usual simplified form an interaction term suitable to cause (52) may be written

$$g_3 \int (N^*P\nu^*\mu^* + P^*N\mu\nu) \, d\Omega. \tag{53}$$

The matrix element corresponding to this interaction is quite similar to those of the previous two interactions and is given by

$$\frac{g_3}{\Omega}. \tag{54}$$

The three constants g_1, g_2, g_3 of (47), (50), and (53) all have the same dimensions, ergs $\times$ cm^3 $\equiv L^5 M T^{-2}$. It is a remarkable fact that their actual values seem to be quite close, namely about 10^{-49} ergs $\times$ cm^3. This fact was noticed independently by Tiomno and Wheeler, Lee, Rosenbluth and Yang, and other authors, and probably is not a coincidence, although at present its significance is unknown. Perhaps it may be correlated to the somewhat similar fact that the electric charges of all the elementary particles are equal. Their common value e plays the role of interaction constant between the electromagnetic field and the various kinds of particles.

It should be noticed further that some of the interactions discussed in the previous sections may not be primary but may be construed as a consequence of others. This possibility, which is exemplified by the Yukawa theory of beta disintegrations, is discussed in Appendix 5.

10. CALCULATION OF TRANSITION RATES

In practical computations it is customary to handle the interaction terms between fields like perturbations. It has been remarked already that this procedure is often quite unsatisfactory because it leads to divergent terms more as a rule than as an exception. Whether ultimately a mathematical procedure will be developed capable of handling the problems in a consistent way or whether more profound modifications of the field theories will be called for is not known at present.

In this elementary discussion we shall avoid the difficulties by terminating the approximation process as soon as a non-vanishing result is achieved, and in some cases by arbitrarily disregarding the contributions of states of very high momentum which are generally responsible for the divergencies.

Extensive use will be made of a general formula[4] which gives the rate at which the transitions from an initial state zero into a continuum of states n take place. (Fig. 1.)

$$R(0 \rightarrow n) = \frac{2\pi}{\hbar} \overline{\mid \mathcal{H}_{no} \mid^2} \frac{dN}{dW}. \qquad (55)$$

$R(0 \rightarrow n)$ is the transition rate, that is the probability of transition per unit time, sometimes improperly referred to as probability of transition. $\mathcal{H}_{no}$ is the matrix element of the perturbation responsible for the transition. It is assumed that it is possible to define the mean square modulus $\overline{\mid \mathcal{H}_{no} \mid^2}$ for a group of states n energetically close to state zero. The expression dN/dW represents the number of final states per unit energy interval for energy close to that of state zero. Strictly speaking, dN/dW will be infinite for a con-

4. See for example Schiff, *Quantum Mechanics* (New York, 1949), p. 193. Notice the non-symmetrical treatment of momentum and energy that has already been discussed in Section 7. The transition discussed here takes place from the state zero to a state n that has about the same unperturbed energy as the state zero. This is in no contradiction with the strict conservation of energy that holds in quantum mechanics as well as in classical mechanics, since the levels of Figure 1 represent unperturbed energies which may differ by slight amounts from the correct energy which is exactly conserved.

tinuum. This infinity, however, is compensated for by the fact that $\mathcal{H}_{no}$ is infinitesimal.

A convenient procedure for avoiding mistakes is to quantize in a finite volume Ω which is later on allowed to increase to ∞. When this procedure is followed the number of final states will be frequently expressible by a formula like (17)

$$dN = \frac{\Omega p^2 \, dp}{2\pi^2 \hbar^3} \tag{56}$$

where p is the momentum of the particle created in the reaction. Both in relativistic and in classical mechanics the energy W and momentum p are related for a single particle of velocity v by

$$dW = v \, dp. \tag{57}$$

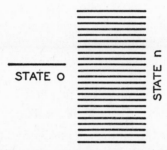

FIG. 1. Transitions from the state zero into the continuum of states n

When a single particle is produced one will have

$$\frac{dN}{dW} = \frac{\Omega p^2}{2\pi^2 \hbar^3 v}. \tag{58}$$

When the reaction yields two particles of equal and opposite momentum, p and $-p$, and speeds v_1 and v_2, formula (56) is unchanged but one finds instead of (57)

$$dW = (v_1 + v_2) \, dp. \tag{59}$$

In the parenthesis the sum of the magnitudes of the velocities and not their vector sum appears. In this case one obtains

$$\frac{dN}{dW} = \frac{\Omega p^2}{2\pi^2 \hbar^3 (v_1 + v_2)}. \tag{60}$$

The case when three particles emerge out of the reaction will be discussed in the special applications.

Formula (55) for the transition rate is obtained from the quantum mechanical perturbation theory in first approximation. According to it no transition should take place unless the matrix element connecting the two states is different from zero. In many important cases one finds, however, that the transition occurs even when the matrix element vanishes. In terms of the perturbation theory this is due to a process of higher approximation. The transition occurs through the intermediary of one or more states connected by non-vanishing matrix elements both to state zero and to the states n. As a rule these intermediate states have energy quite different from state zero, and therefore no permanent transition of the system into them is possible on account of the conservation of energy.

One applies in such cases the standard rules of the perturbation theory[5] (second approximation). The transitions from zero to n occur as if the two states were connected directly by an "effective matrix element":

$$\mathcal{H}'_{no} = \sum_m \frac{\mathcal{H}_{nm}\,\mathcal{H}_{mo}}{W_o - W_m} \tag{61}$$

The sum is extended to all the intermediate states m connected to states zero and n by non-vanishing matrix elements. In practice such sums are frequently divergent. When this is the case we shall adopt crude cut-off procedures in order to obtain so called "plausible" results.

11. DEVELOPMENT PARAMETERS

A well-advertised fact is that the fine-structure constant $e^2/\hbar c = 1/137$ plays the role of development parameter in radiative phenomena. This means that in a series expansion in powers of e obtained when applying the perturbation method successive terms decrease in order of magnitude as successive powers of $\sqrt{e^2/\hbar c} = .085$. Actually, while this is true dimensionally, infinite numerical coefficients often are encountered which

5. See for example: Schiff, *Quantum Mechanics*, p. 196.

obviously spoil the argument. In any case, the small value of the development parameter may be taken as an indication of the reliability of the approximation procedure.

As an example of the meaning of the development parameters appropriate to the various forms of field interactions we shall discuss first the Yukawa interaction (41). The state representing a proton P at rest is connected by a non-vanishing matrix element (42) to a state $(N + \Pi)_p$ representing a positive pion and a neutron with equal and opposite momenta p and $-p$. From the perturbation theory we know that the unperturbed state (P) will be mixed with states $(N + \Pi)_p$ in the linear combination

$$(P) + \sum_p \frac{\mathcal{H}_{po}}{\Delta W} (N + \Pi)_p \tag{62}$$

where ΔW is the energy difference between the two states. In this approximation the probability of finding the system in the state $(N + \Pi)_p$ is $|\mathcal{H}_{po}|^2/\Delta W^2$ and the probability of finding the system in any one of the states $(N + \Pi)_p$ is $\Sigma_p |\mathcal{H}_{po}|^2/\Delta W^2$ where the sum should be taken for all allowable values of p. In order to evaluate this sum one replaces the matrix element $\mathcal{H}_{po}$ by its value (42) $e_2\hbar c/\sqrt{2\Omega w}$ where $w = \sqrt{\mu^2 c^4 + c^2 p^2}$ is the energy of the pion. Since the pion mass is small compared to that of the nucleon the energy difference between the state $(N + \Pi)_p$ and the unperturbed state (P) can be taken equal to w. The sum can then be converted into an integral by multiplying the summand by the expression (56) of the number of states between p and $p + dp$ and integrating to all values of p. Introducing as a variable instead of p the new variable $\eta = p/\mu c$, one finds

$$\Sigma \frac{|\mathcal{H}_{po}|^2}{\Delta W^2} = \frac{e_2^2}{4\pi\hbar c} \int_0^\infty \frac{\eta^2 \, d\eta}{\pi(1 + \eta^2)^{3/2}}.$$

Unfortunately the integral has a logarithmic divergence at $\eta = \infty$. We get around this difficulty by one of the crude cut-off procedures already mentioned. Instead of integrating up to infinite values of the momentum we limit the integral to maximum momentum of the order μc, that is, to values of η of order one. A justification for this cut-off is the expectation that the expression (42) of the matrix element may in fact break down at relativistic velocities of the pion. The integral in the formula above becomes now of the order of magnitude of unity and the expression itself becomes of the order of magnitude

$$\frac{e_2^2}{4\pi\hbar c} \tag{63}$$

which is the usual expression adopted for the development parameter of the Yukawa theory. The development parameter represents therefore as order of magnitude the fraction of the states $(N + \Pi)$ mixed in with (P). The expression (63) is the analogue of the fine-structure constant of the radiation theory. The factor 4π in the denominator is there because in the Yukawa theory one adopts rationalized units. The development parameter (63) is not a small number. It is difficult to assign to it a precise value since no meson theory accounts for the nuclear forces well enough to permit a determination of the coupling constant. However, it is believed that (63) may be of the order of $1/4$. Then e_2 would be about twenty times the elementary electric charge.

These large values of e_2 and of (63) mean of course that a physical proton is rather poorly represented by the mathematically simple state that we have indicated by (P). Actually, it is a mixture of the state (P) representing a proton proper, and of states $N + \Pi$ representing a neutron with a positive pion nearby. The physical proton is therefore far from elementary. The usual way of describing the situation is to say that a proton occasionally emits a positive pion and converts temporarily into a neutron. Soon afterward it reabsorbs the pion and goes back to the proton form. The neutron has a similar behavior. The probability of finding the proton in the state $N + \Pi$ rather than in the state (P) is quite large, being of the order of magnitude of the development parameter, or perhaps 25 per cent. When two nucleons approach each other their surrounding pion fields interact in a manner that gives rise to the strong short-range forces between nucleons as will be explained in Section 17.

The development parameters for the other types of interaction are obtained in a similar way. For the interaction (44) the parameter is given by an expression like (63), namely

$$\frac{e_3^2}{4\pi\hbar c}. \tag{64}$$

Since e_3 turns out to be very small this parameter is also quite small so that the perturbation method is presumably reliable. The same is true of the development parameters for the interactions (47), (50), (53). These parameters are given by

$$\frac{g^2 m^4 c^2}{\hbar^6} \tag{65}$$

where g should be replaced in the three cases by g_1, g_2, g_3 and m should be replaced for the cases (47), (50) by the mass of the electron and for case (53) by that of the muon. In all cases the parameter (65) is quite small, ranging in value from 10^{-22} to 10^{-12}. The approximation method for these cases should therefore be reliable provided the actual form of the interaction terms has been guessed correctly.

CHAPTER THREE

The Interaction Constants

12. ELECTROMAGNETIC AND YUKAWA INTERACTION CONSTANTS

In the preceding chapter six interaction processes have been discussed. They do not cover all possibilities. There could be additional interactions among the elementary particles, and besides there are particles whose existence is either known or suspected which we have left out of consideration because too little is known of their properties. For each of the six interaction processes of Chapter II a constant has been introduced that determines its strength. Three of them have the dimensions of an electric charge and three have the dimensions of energy × volume. The first three are

e—the elementary electric charge that determines the strength of the electromagnetic interaction.

e_2—the interaction constant of the Yukawa theory determining the strength of the interaction between pions and nucleons.

e_3—the constant of an interaction that has been postulated to act between pions, muons, and neutrinos, which could be responsible for the spontaneous decay of the pion.

The three constants with dimensions energy × volume are

g_1—the interaction constant of the beta processes.

g_2—an interaction that has been postulated to act between muons, electrons, and neutrinos and which could be responsible for the spontaneous decay of the muon.

36

g_3—the interaction constant of a hypothetical process similar to the beta interaction except that the electron is replaced by a muon.

Perhaps future developments of the theory will enable us to understand the reasons for the existence and the strength of these various interactions. At present, however, we must take an empirical approach and determine the values of the various constants from the intensity of the phenomena that are caused by them. In Appendix 5 some of the possible relationships between various constants are discussed.

The electromagnetic interaction constant e is of course very well known. Its value is

$$e = 4.8025 \times 10^{-10} \quad cm^{3/2} \, gr^{1/2} \, sec^{-1}.$$

It is a remarkable fact that the value of the elementary charge is the same for all charged particles: electrons, protons, muons, and pions. This identity is presumably related to the conservation of electricity and to the fact that particles can change into each other. For example, the beta processes in which a neutron changes into a proton, an electron, and a neutrino would be incompatible with the conservation of electricity if the absolute values of the charge of the proton and the electron were different. We will not discuss the electromagnetic processes here except for a few cases in which they are necessary in order to understand other properties of the elementary particles.

The Yukawa interaction is responsible for the nuclear forces, and the value of the corresponding interaction constant e_2 should be adjusted to fit the experimental values of these forces. A sketch of the Yukawa theory of nuclear forces will be given later in Section 17. It will be found there that in order to obtain an order-of-magnitude agreement between the observed and calculated strength of the nuclear forces the constant e_2 must be taken about twenty times larger than the elementary electric charge. In the numerical calculations we shall adopt the value

$$e_2 = 10^{-8} \, cm^{3/2} \, gr^{1/2} \, sec^{-1}. \tag{66}$$

It should be stressed that this value is merely an order of magnitude. The quantitative agreement between the Yukawa theory of nuclear forces and experiment is so indefinite that it has not been possible to determine the precise form of the interaction energy and therefore the value of the constant must also be left uncertain. In particular, the value of the spin of the pion is not known. It could be 0 or 1, or perhaps have an even higher value. In our simplified discussion a spin zero has always been assumed.

13. DECAY OF THE PION AS DIRECT PROCESS

The spontaneous conversion of the pion into a muon and a neutrino discussed in Section 9 may be due to the direct effect of the interaction (44). The lifetime for this process will now be calculated on this assumption, in order to deduce from the comparison of theory and experiment the value of the interaction constant e_3.

A pion at rest has rest energy μc^2. This energy is converted into the energy of the neutrino of momentum p and the muon of momentum $-p$. Conservation of energy requires

$$cp + \sqrt{\mu_1^2 c^4 + c^2 p^2} = \mu c^2 \qquad (67)$$

where μ_1 is the muon mass. From this equation one computes the momentum p of the two particles produced in the reaction. Let τ_π be the lifetime of the pion. The rate of the pion decay according to the reaction $\Pi \to \mu + \nu$ is $1/\tau_\pi$ and can be calculated with (55) and (60) using the matrix element (45). Since a pion at rest disappears, its energy w_s in (45) will be μc^2. one finds

$$\frac{1}{\tau_\pi} = \frac{2\pi}{\hbar} \cdot \left(\frac{e_3 \hbar c}{\sqrt{2\Omega\mu c^2}}\right)^2 \cdot \frac{\Omega p^2}{2\pi^2 \hbar^3 (v_1 + v_2)} = \frac{e_3^2 p^2}{2\pi\hbar^2\mu(v_1 + v_2)}.$$

All spins have been neglected, and v_1 and v_2 are the speeds of the neutrino and of the muon. One has therefore $v_1 = c$. The velocity v_2 of the muon and the momentum p can be computed from the conservation of energy. Assuming $\mu = 276\,m$ and $\mu_1 = 210\,m$, with m the electron mass, one finds by solving (67) $p = 58.1\,mc$, from which it follows that the neutrino carries away about 30 Mev while the kinetic energy of the muon is only 4.1 Mev. The velocity of the muon is $0.27\,c$. With these numerical values one finds

$$1/\tau_\pi = 3.8 \times 10^{37}\, e_3^2. \qquad (68)$$

From the experimental value $\tau_\pi = 2 \times 10^{-8}$ seconds given by Martinelli and Panofski we obtain the interaction constant

$$e_3 = 1.1 \times 10^{-15} = 2.4 \times 10^{-6} e. \tag{69}$$

It will be seen in Appendix 5 that this interaction may not be primary. It may be an indirect effect of other interactions that have been postulated.

14. THE BETA INTERACTION

The interaction responsible for the beta transitions has already been given in a crude form in Section 9. Next to the case of the radiation theory this is the process on which we have the greatest amount of experimental information. In spite of this favorable situation the attempts to substitute for (47) a more detailed expression have not led so far to a final conclusion. Most of the attempts are based on assuming an interaction term in the Lagrangian formed with expressions patterned on those given in Appendix 4. Since the term in the Lagrangian must be a scalar, one might for example proceed as follows: The expression $\tilde{P}\beta N$ is a scalar according to (2) in Appendix 4; so is the expression $\tilde{e}\beta\nu$. Therefore their product multiplied by an arbitrary interaction constant $-g_\beta$ will be an invariant scalar that could be added to the Lagrangian to represent the interaction. In passing from the Lagrangian to the Hamiltonian there is in this case only a change of sign and therefore a possible invariant interaction term would be

$$g_\beta \int (\tilde{P}\beta N\tilde{e}\beta\nu + \tilde{N}\beta P\tilde{\nu}\beta e) \, d\Omega. \tag{70}$$

In writing this expression the complex-conjugate quantity has been added since the Hamiltonian must be Hermitian. The interaction (70) is called scalar interaction. One can form four additional relativistic invariant types of interaction by using the expressions (3) to (6) of Appendix 4 in a similar manner. They are called respectively pseudoscalar, vector, pseudovector and tensor interactions. For example, the pseudovector interaction would lead to a term of the following form:

$$g_{ps} \int d\Omega \left\{ \tilde{P} \frac{\alpha \times \alpha}{i} N \cdot \tilde{e} \frac{\alpha \times \alpha}{i} \nu - \tilde{P} \frac{\alpha_1\alpha_2\alpha_3}{i} N\tilde{e} \frac{\alpha_1\alpha_2\alpha_3}{i} \nu \right. \\ \left. + \text{compl. conj.} \right\} \tag{71}$$

where g_{ps} represents the value of the interaction constant for this case. Also, linear combinations of these various interactions may be adopted. Attempts have been made to compare with experiment the consequences of the various interaction forms in order to decide which if any of them gives acceptable results. It turns out, unfortunately, that for the so-called permitted transitions the shape of the beta-ray spectrum is the same for all types of interaction. The excellent agreement between the theoretical and experimental shape of the permitted spectra therefore does not offer a clue for a decision. On the other hand, the different types of interaction give different selection rules. For example, the scalar and the vector interactions give as a selection rule for permitted transitions:

No change of nuclear spin, no change of parity. (F-selection rules.)

The selection rules for the pseudovector or tensor interactions are instead:

Change of spin by 0 or ± 1, no change of parity. $0 \to 0$ transitions forbidden. (GT-selection rules.)

It appears that experimental evidence tends to favor the GT-selection rules, or perhaps a mixture of the F and GT rules. This evidence, however, is not quite conclusive, since unfortunately little is known experimentally of the spin and parity of beta-active substances and one is usually forced to guess on the basis of nuclear models which, although plausible, are not very well established. If one assumes the GT rules one would be led to use either the pseudovector interaction (71) or the tensor interaction. In most cases these two interactions give almost identical results.

Wigner and Critchfield have considered in addition the possibility of a linear combination of scalar, pseudovector and pseudoscalar interactions which can be expressed in a remarkably symmetrical form. Its results show an agreement with experiment no worse than that of other theories. The results of the comparison of theory with experiment can be summarized by saying that the shapes of the permitted spectra and of several forbidden spectra seem to follow the theoretical shape or shapes with surprising accuracy. On the other hand, only indifferent success has been met until now in the attempts to correlate the intensity of the beta transitions with the matrix elements calculated from nuclear models. Of course, the nuclear models used are very crude so that the test is not very conclusive.

The various types of interaction lead to different predictions as to the angular correlations of the directions in which the electron and the neutrino are emitted. Since, however, the observation of the recoil of the neutrino is not sufficiently accurate it has not proved possible until now to draw reliable conclusions. In the present discussion we shall limit ourselves to the use of the simplified interaction (47) and to the derivation of its most immediate consequences.

In calculating the matrix element of a beta transition one cannot apply without modification the rules of Section 7 because the neutron and the proton cannot be represented by plane waves but are bound by strong forces inside the nucleus. For this same reason no conservation of momentum holds for the four particles in equation (46) since the residual nucleus can take up any amount of momentum.

The calculation will be simplified by disregarding the Coulomb forces acting on the electron. Also the usual assumption will be made that the wave lengths of the electron and the neutrino are large compared to the nuclear dimensions and therefore the phase factor can be omitted in computing the contribution to the matrix element of the factors e, e^*, v, v^* according to (21). Each of these expressions contributes then the constant factor $1/\sqrt{\Omega}$, except for the sign which is of no relevance. The first term of (47) is the one that is responsible for the beta processes with emission of negative electrons, while the second term produces the reverse transition and the beta processes with positron emission.

Since proton and neutron are bound in the nucleus their contribution to the matrix element of (47) should be computed according to (22) and not (21). Denoting by N and P the wave functions of the neutron that is present in the nucleus before the transformation and of the proton that is there afterward, one finds the matrix element

$$\mathcal{H}_{n0} = \frac{g_1}{\Omega} \int P^* N \, d\Omega = \frac{g_1}{\Omega} \mathfrak{M}. \tag{72}$$

This matrix element can be used for calculating the probability of the transition from the initial state to a state in which an electron is emitted with momentum between p and $p + dp$ and the neutrino escapes with momentum q, carrying away the balance of the available energy. If p_0 is the maximum momentum with which electrons can be emitted, the neutrino energy cq will be the difference between the energy of electrons with momenta[1] p_0 and p:

$$cq = \sqrt{m^2c^4 + c^2p_0^2} - \sqrt{m^2c^4 + c^2p^2}. \tag{73}$$

We compute the probability of transition with formula (55). The term dN/dW will be computed with (58), substituting however for p the momentum q of the neutrino and for v its velocity, c. An additional factor representing the number of states of an electron of momentum between p and $p + dp$ given by (56) must be included. The final result is

$$\frac{2\pi}{\hbar} \left| \frac{g_1 \mathfrak{M}}{\Omega} \right|^2 \frac{\Omega q^2}{2\pi^2\hbar^3 c} \frac{\Omega p^2 \, dp}{2\pi^2\hbar^3} = \frac{g_1^2 |\mathfrak{M}|^2}{2\pi^3\hbar^7 c} q^2 p^2 \, dp$$

$$= \frac{g_1^2 |\mathfrak{M}|^2}{2\pi^3\hbar^7 c^3} \left(\sqrt{m^2c^4 + c^2p_0^2} - \sqrt{m^2c^4 + c^2p^2} \right)^2 p^2 \, dp. \tag{74}$$

This formula gives the shape of the spectrum of the emitted beta rays. From it one can also calculate the total probability of transition for the emission of electrons of any allowable energy by integrating from zero to p_0. Introducing for convenience the new variables η and η_0 representing the momenta in units mc one finds the lifetime for the transition given by the formula

$$\frac{1}{\tau} = \frac{g_1^2 |\mathfrak{M}|^2 m^5 c^4}{2\pi^3\hbar^7} F(\eta_0) \tag{75}$$

where

$$F(\eta_0) = \int_0^{\eta_0} \left(\sqrt{1 + \eta_0^2} - \sqrt{1 + \eta^2} \right)^2 \eta^2 \, d\eta$$

$$= -\frac{\eta_0}{4} - \frac{\eta_0^3}{12} + \frac{\eta_0^5}{30} + \frac{1}{4} \sqrt{1 + \eta_0^2} \ln \left(\eta_0 + \sqrt{1 + \eta_0^2} \right). \tag{76}$$

The term $|\mathfrak{M}|^2$ in (75) is related to the nuclear wave functions by (72). Its calculation requires therefore knowledge of the wave functions of the initial and final state of the nucleus. Since these wave functions in most cases are not known, it is usually impossible

---

1. Since electron and neutrino are relativistic particles the relativistic formulas (25) should be used rather than (21). One finds, however, that the shape of the beta spectrum is given correctly by the logically inconsistent mixed use of relativistic and nonrelativistic formulas used in the text.

to do much more than guess the order of magnitude of $\mathfrak{M}$. There are some favorable cases for which one has reason to believe that the neutron and the proton have approximately equal wave functions. Such cases are frequently encountered for light elements, particularly for the so-called Wigner nuclei. If P and N in (72) are the same wave functions, the integral reduces to the normalization integral and is therefore equal to one. In such cases, therefore, the simplified theory leads to the simple result $\mathfrak{M} = 1$. Somewhat more complicated results are expected, however, even in this simplest case if one of the more elaborate forms of interactions that have already been discussed is adopted. In particular, if one uses the GT-selection rules, one finds for $\mathfrak{M}$ the following expression instead of (72):

$$\mathfrak{M} = \int P^* \sigma N \, d\Omega \qquad \text{(for } GT \text{ rules)} \qquad (77)$$

where σ is the Pauli spin operator of the nucleons. Even with (77), however, $\mathfrak{M}$ usually is of the order of magnitude of at least 1 in the simplest cases.

An additional complication concerns the shape of the spectrum and the form of the function $F(\eta_0)$. The action of the Coulomb field of the nucleus on the electron has been neglected so far. The Coulomb field distorts the electron wave functions and changes the shape of the spectrum especially at low electron energies. Similarly, the function $F(\eta_0)$ will be more complicated. The reader is referred to specialized articles for a more elaborate discussion of the beta-ray theory and of its comparison with experiment.

The value of the interaction constant that gives the best agreement with experimental data is for all forms of interaction of the order of 10^{-49} ergs $\times$ cm^3. The precise value depends of course on the interaction form that is assumed. For the pseudoscalar interaction (71) one finds for example

$$g_1 = 2 \times 10^{-49} \text{ ergs } \times \text{ cm}^3. \qquad (78)$$

g_1 has the dimensions of energy $\times$ volume. In order to appreciate its order of magnitude we divide it by the classical volume of the electron:

$$\frac{4\pi}{3} \left(\frac{e^2}{mc^2} \right)^3 = 9.4 \times 10^{-38} \text{ cm.}^3.$$

The result is 2.1×10^{-12} ergs or about 1.3 ev. This may give an idea of how weak the beta interaction is. For example, the interaction potential between two nucleons which also is sometimes represented by a potential hole of radius equal to the classical electron radius has a depth of 20 Mev, that is, about ten million times greater.

15. SPONTANEOUS DECAY OF THE MUON

The spontaneous decay of the muon with emission of an electron has been attributed in Section 9 to a process $\mu \rightarrow e + 2\nu$ due to an interaction term (50) with matrix element g_2/Ω given by (51).

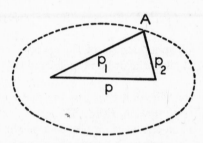

FIG. 2. Momenta of the electron (p) and of the two neutrinos (p_1 and p_2)

The calculation of the lifetime of the muon due to this interaction is a little bit more complicated than the similar calculation of the lifetime of the pion carried out in Section 13 because in this case three particles instead of two are produced. From this it follows that the electron resulting from the disintegration of a muon at rest will have a continuous energy spectrum. The maximum electron energy is obtained when the two neutrinos fly out in the same direction and it is one-half the mass energy of the muon or about 53 Mev if one neglects the rest mass of the electron. In the calculation that follows this will be done.

When a muon at rest decays into an electron and two neutrinos the vector sum of the momenta p, p_1, p_2 of these three particles will vanish.

The rate of transition into a state in which the electron escapes with momentum between p and $p + dp$ is obtained by multiplying the number of these electron states given by (56) by the rate of transition into one electron state computed with (55). Neglecting the spin of the particles, the number of electron states is given by $\Omega p^2 \, dp/(2\pi^2\hbar^3)$. With the expression g_2/Ω of the matrix element, one finds in this way the transition rate

$$\frac{2\pi}{\hbar} \left(\frac{g_2}{\Omega}\right)^2 \frac{\Omega p^2 \, dp}{2\pi^2\hbar^3} \frac{dN}{dW}. \tag{79}$$

The number of neutrino states per unit energy interval, dN/dW, still must be calculated. In Figure 2 are represented the momentum p of the electron and the momenta p_1 and p_2 of the two neutrinos. Since the vector sum vanishes the three vectors form the sides of a triangle. The rest mass of the electron being neglected, the energy of the three emitted particles is

$$W = c(|\, p_1 \,| + |\, p_2 \,| + |\, p \,|). \tag{80}$$

If we keep the momentum p of the electron and the energy W constant, the sum of the magnitudes of p_1 and p_2,

$$|\, p_1 \,| + |\, p_2 \,| = W/c - |\, p \,|,$$

will also be constant. From this it follows that the vertex A of the triangle will move on an ellipsoid. By simple geometrical considerations one can calculate its volume:

$$\frac{\pi}{6}\left(\frac{W^3}{c^3} - 3\frac{W^2}{c^2}p + 2\frac{W}{c}.p^2\right).$$

The corresponding phase-space volume is the product of the above expression and the volume Ω. The number N of neutrino states up to the energy W is the phase-space volume divided by $(2\pi\hbar)^3$. By differentiating this expression with respect to W, dN/dW is obtained. In it we put finally $W = \mu_1 c^2$ where μ_1 is the muon mass and we find

$$\frac{dN}{dW} = \frac{\Omega c\mu_1^2}{48\pi^2\hbar^3}\left(3 - 6\frac{p}{\mu_1 c} + 2\frac{p^2}{\mu_1^2 c^2}\right). \tag{81}$$

Substituting in (79) one obtains the rate of transition

$$\frac{g_2^2\mu_1^2 c}{48\pi^3\hbar^7}\left(3 - \frac{6p}{\mu_1 c} + \frac{2p^2}{\mu_1^2 c^2}\right)p^2 \, dp. \tag{82}$$

The total rate of transition, that is, the inverse of the lifetime of the muon, is obtained by integrating (82) to all values of p, from 0 to $\mu_1 c/2$. The result is

$$\frac{1}{\tau_\mu} = \frac{7}{7680\pi^3} \frac{g_2^2 \mu_1^5 c^4}{\hbar^7}.$$

(83)

Experimentally $\tau_\mu = 2.15 \times 10^{-6}$ seconds. Substituting this

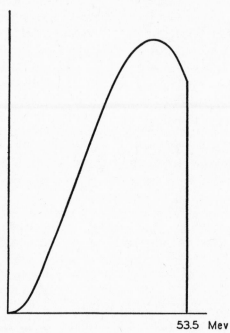

53.5 Mev

FIG. 3. Energy spectrum of the electrons from the spontaneous muon decay.

value and $\mu_1 = 210m$ in (83) one obtains the value of the coupling constant, g_2 :

$$g_2 = 3.3 \times 10^{-49}.$$

(84)

The closeness of the values of g_1 and g_2 is very striking.

Formula (82) gives the spectral distribution of the emitted electrons. Its results are represented graphically in Figure 3. Experimentally the shape of the spectrum of these electrons is known only very crudely. The theoretical spectrum, however, is not incompatible with the measurements of Steinberger and Anderson. It should be remembered that the precise form of the interaction (50) is not known. Various types of interaction could be used in this case as we have already seen for the beta interaction. Wheeler has made an extensive study of the effect that a specific choice of the interaction form has on the shape of the spectrum and on the other features of disintegration of the muon.

16. FORCED DECAY OF THE MUON

The muon may decay also by a different process which is called forced decay because it happens only when the muon is captured near a nucleus. The forced decay is observed only for negative muons, presumably because only muons of this sign are attracted near the nuclei by electric forces. In Section 9 forced decay has been attributed to the reaction (52) with matrix elements (54).

In order to estimate the value of the coupling constant g_3 we shall assume that when a negative muon comes to rest inside a material it is very rapidly captured in the vicinity of one of the nuclei in an orbit which is the analogue of the lowest Bohr orbit, only much smaller. Its radius will be obtained by dividing the Bohr radius $a = 5.3 \times 10^{-9}$ cm by a factor Z due to the nuclear charge and by a factor 210, the ratio of the muon and the electron masses. The radius of this Bohr-like orbit is therefore

$$a' = a/(210Z) = 2.5 \times 10^{-11}/Z \text{ cm}. \qquad (85)$$

The time required for the capture in this orbit is estimated to be of the order of 10^{-13} seconds so that the probability that the muon may decay spontaneously during this time is negligible. After the capture has taken place the destruction of the muon may occur either by the spontaneous process of the previous section or by the forced process. The relative probability of these two mechanisms depends on the nucleus that has captured the muon since the probability of forced decay increases rapidly with Z for two reasons. For large Z the radius a' of the orbit is smaller so that the muon is found on the average closer to the nucleus. The nucleus furthermore contains a large number of protons that may induce the decay. The experimental results confirm this general picture and indicate that the process (52) is negligible for very light capturing elements and is dominant for heavier elements. According to the measurements of Ticho the rate of transition for the two competing processes is approximately equal for $Z = 11$ (sodium). Since the spontaneous decay has a lifetime of 2.15×10^{-6} seconds we conclude that the rate of forced decay of a muon captured by a nucleus with $Z = 11$ is also

$$1/(2.15 \times 10^{-6}) \text{ sec}^{-1} \qquad \text{(for } Z = 11).$$

We first calculate the rate of transition for the process (52) when only one muon and one proton of zero velocity are present in a volume Ω. The matrix element according to (54) is g_3/Ω. The rate of transition can then be calculated with (55) and (60):

$$\text{rate of transition} = \frac{2\pi}{\hbar} \left(\frac{g_3}{\Omega}\right)^2 \frac{\Omega p^2}{2\pi^2 \hbar^3 (c + v_N)}$$

$$= \frac{g_3^2}{\pi \hbar^4} \frac{p^2}{c + v_N} \frac{1}{\Omega}. \tag{86}$$

The term $c + v_N$ that appears in the denominator is the sum of the speeds c of the neutrino and v_N of the neutron. p is the magnitude of the equal and opposite momenta of N and ν. p and v_N can be computed from the conservation of energy. If the small mass difference between proton and neutron is neglected, the available energy is the rest energy of the muon, $210mc^2$. This energy appears as kinetic energy of the neutrino and of the neutron, both having a momentum p. Therefore

$$210mc^2 = cp + \frac{p^2}{2M}. \tag{87}$$

M = nucleon mass. From this equation one finds the momentum p of the two particles, $p = 199mc$. The velocity of the neutron given by this momentum divided by the neutron mass is $v_N = 0.11c$. From (86) one finds now

$$\text{rate of transition} = 11{,}400 \frac{g_3^2 m^2 c}{\hbar^4 \Omega}. \tag{88}$$

The rate of transition is inversely proportional to Ω because we have assumed that there is one proton in the volume Ω and therefore the concentration of protons is inversely proportional to Ω.

In calculating the probability that a muon captured in a Bohr-like orbit near a nucleus of charge Z is destroyed by the forced decay process we use (88) with the following changes: (a) multiply by Z because there are Z protons that can effect the capture; (b) substitute for Ω the effective volume of the Bohr-like orbit of the muon of radius a', namely $\pi a'^3$; (c) make a further change in the coefficient of (88) in order to account for the possibility that the process may be inhibited by the Pauli principle. In the forced decay a rather low-energy neutron is produced and the reaction will be possible only if the orbit into which the neutron may be created is not occupied in the nucleus. The reduction factor would be difficult to calculate accurately, but we estimate that a reduction of the coefficient from 11,400 to 5,000 may be of the correct order of magnitude.

In conclusion, the probability of transition for the forced decay is

$$\frac{1}{\tau_3} = \frac{5000(210)^3}{\pi}\, g_3^2\, \frac{m^2 c Z^4}{\hbar^4 a^3}. \tag{89}$$

Notice the proportionality to Z^4 which bears out the great difference in the intensity of this process for light and heavy nuclei.

From the experimental fact previously quoted that, for $Z = 11$, τ_3 is about equal to 2.15×10^{-6} seconds, one calculates the value of the coupling constant g_3 :

$$g_3 = 1.3 \times 10^{-49} \tag{90}$$

It has already been noticed that the three coupling constants, g_1, g_2, g_3 are quite close: $g_1 = 2 \times 10^{-49}$, $g_2 = 3.3 \times 10^{-49}$, $g_3 = 1.3 \times 10^{-49}$. This similarity probably is not an accidental coincidence but has some deep meaning which, however, is not understood at present.

CHAPTER FOUR

Pions, Nucleons, and Anti-Nucleons

17. YUKAWA THEORY OF NUCLEAR FORCES

The Yukawa theory of nuclear forces is based on the assumption that the forces acting between two nucleons are transmitted by the pion field that surrounds them. The phenomenon is usually described by stating that when two nucleons approach each other one of them may emit a pion which is absorbed by the other. Assuming for example that the two nucleons are a proton and a neutron, it may happen first that the proton emits a positive pion so that an intermediate state with two neutrons and a positive pion will be formed. The positive pion is subsequently absorbed by the original neutron and converted into a proton according to the reaction

$$P + N \rightarrow N' + N + \Pi^+ \rightarrow N' + P'. \tag{91}$$

The overall result of (91) is the scattering of the two nucleons which change their momentum direction. This scattering is interpreted as if it were due to a force acting between the two nucleons.

Formally the scattering (91) is a second approximation process. The Yukawa interaction (41) has no matrix elements connecting the initial to the final state of (91). These two states, however, are both connected to the intermediate state $N' + N + \Pi^+$. The energy of this intermediate state is greater than that of the initial state. Indeed, if we assume that the nucleons are slow and their kinetic energy is negligible, the difference in energy between the intermediate and the initial state will be the pion energy

$$w = \sqrt{\mu^2 c^4 + c^2 p^2} \tag{92}$$

Transitions will occur in second approximation between the initial and final state of (91). Such transitions will appear phenomenologically to be caused by a direct interaction between the neutron and the proton. In order to compute this interaction we use the two matrix elements (42) corresponding to the two Yukawa steps of (91). The contribution of the intermediate state of (91) to the apparent matrix element between initial and final state according to (61) is the product of these two matrix elements divided by the energy difference, $-w$, between initial and intermediate states. The final result, however, should be multiplied by two because an equal contribution is given by another intermediate state, $P + P' + \Pi^-$, for which a negative pion is emitted by the neutron. The possible contribution of intermediate states in which a neutral pion is emitted will be disregarded at this time. In conclusion, the effective matrix element is

$$\mathcal{K}'_{no} = -\frac{2}{w}\left(\frac{e_2\hbar c}{\sqrt{2\Omega w}}\right)^2 = -\frac{e_2^2\hbar^2 c^2}{\Omega}\frac{1}{\mu^2 c^4 + c^2 p^2}. \tag{93}$$

The pion momentum p is the difference between the momenta of the original proton and of the final neutron since momentum is conserved in the first step of (91). It is also equal to the difference between the momenta of the final proton and the initial neutron. These two differences, of course, are equal because momentum is conserved between the initial and final state of (91).

$$p = p_P - p_{N'} = p_{P'} - p_N. \tag{94}$$

Phenomenologically the transition from initial to final state of (91) could be described as due to a force acting between the proton and the neutron without explicit reference to the role played by the emission and absorption of the pion. This force, however, is represented not by an ordinary potential but by a so-called exchange potential. In simple terms this can be understood by observing that in the emission of the positive pion the original proton changes to a neutron while in the subsequent absorption of the positive pion the original neutron changes to a proton so that the role of the two nucleons is interchanged .

We will assume therefore that between proton and neutron there exists a mutual energy represented by

$$V = U(r)P_{ex} \tag{95}$$

where P_{ex} is the neutron-proton exchange operator.[1] We propose to determine the function $U(r)$ in such a way that the matrix elements of (95) become identical to (93). The initial and the final states of (95) are described by the wave functions

$$\psi_0(r_P, r_N) = \frac{1}{\sqrt{\Omega}} e^{(i/\hbar) p_P \cdot r_P} \times \frac{1}{\sqrt{\Omega}} e^{(i/\hbar) p_N \cdot r_N}$$

$$\psi_n(r_P, r_N) = \frac{1}{\sqrt{\Omega}} e^{(i/\hbar) p_{P'} \cdot r_P} \times \frac{1}{\sqrt{\Omega}} e^{(i/\hbar) p_{N'} \cdot r_N}.$$

The matrix element of the interaction (95) is therefore

$$V_{n0} = \frac{1}{\Omega^2} \iint U(r_N - r_P) \, e^{(i/\hbar)\{(p_{P'}-p_N)\cdot r_P + (p_{N'}-p_P)\cdot r_N\}} \, d^3 r_P \, d^3 r_N.$$

Notice the effect of the exchange operator P_{ex} in the exponent. When the new integration variables r_P and $r = r_N - r_P$ are introduced in place of r_N, r_P the above formula becomes

$$V_{n0} = \frac{1}{\Omega^2} \left(\int U(r) e^{-(i/\hbar)(p_P - p_{N'})\cdot r} \, d^3 r \right)$$
$$\int e^{(i/\hbar)(p_{P'} - p_N + p_{N'} - p_P)\cdot r_P} \, d^3 r_P. \tag{96}$$

The last integral vanishes in all cases except when the momentum is conserved, that is, when

$$p_P + p_N = p_{P'} + p_{N'}.$$

If this is the case, the value of the second integral in (96) is the normalization volume Ω. We have therefore, using (94),

$$V_{n0} = \frac{1}{\Omega} \int U(r) e^{-(i/\hbar) p \cdot r} \, d^3 r. \tag{97}$$

We want to determine $U(r)$ so that (97) and (93) become identical. That is

$$\int U(r) \, e^{-(i/\hbar) p \cdot r} \, d^3 r = - \frac{e_2^2 \hbar^2 c^2}{\mu^2 c^4 + c^2 p^2}. \tag{98}$$

1. P_{ex} is an operator which can be applied to functions of the coordinate of a neutron and a proton. The result is the same function in which, however, the coordinates of the neutron and the proton are interchanged. Its operational definition is

$$P_{ex}\psi(x_P, x_N) = \psi(x_N, x_P).$$

This equation gives the Fourier components of $U(r)$. In order to determine the function U one inverts the Fourier transformation according to the standard rules. The result is

$$U(r) = -\frac{e_2^2}{8\pi^3\hbar} \int \frac{e^{(i/\hbar)p\cdot r}\, d^3p}{\mu^2c^2 + p^2} = -\frac{e_2^2}{2\pi r}\, e^{-(uc/\hbar)r}. \qquad (99)$$

This formula gives the famous expression of the Yukawa potential with its characteristic range equal to the Compton wave length for the pion:

$$\frac{\hbar}{\mu c} = 1.4 \times 10^{-13} \text{ cm.}$$

This important result may be readily understood qualitatively with the uncertainty principle. The force field of a nucleon extends as far as the pions which it continuously emits and reabsorbs are able to reach. When a pion is emitted the system "borrows" an energy of the order of μc^2. This loan, according to the time-energy complementarity, cannot last a longer time than about $\hbar/\mu c^2$. In this time the pion, even if it travels with the velocity of light, cannot move farther than $\hbar/\mu c$. This is, therefore, the order of magnitude of the range.

The simplified outline of the theory of the Yukawa forces between two nucleons given here is unsatisfactory in many respects. The interaction (41) that has been assumed between the nucleon and pion fields not only does not satisfy relativistic invariance but disregards also the spin properties of the nucleons. It should be replaced by other forms of interaction like perhaps (39) or (40). As long as (41) is used there is of course no way of explaining theoretically spin-dependent nuclear forces.

The Yukawa theory of nuclear forces has been carried out on the basis of various assumptions as to the spin and parity properties of the pion and the type of coupling between pion and nucleon fields. In particular, the case of a pseudoscalar pion and of a vector pion, that is, of a pion with spin one, have been investigated extensively. Qualitatively both theories lead to spin-dependent nuclear force and to quadrupole forces. Also the contribution of neutral pions to the nuclear forces has been taken into account. The Yukawa theory with charged pions gives exchange forces. Neutral pions yield instead ordinary forces. Qualitatively therefore the Yukawa theory is capable of explaining the existence of the various types of forces that are known to act between nucleons.

It has not been possible, however, to construct a theory that is quantitatively satisfactory. The chief difficulty is that some terms in the potentials become too strongly attractive at short distances to permit the building up of deuteron states with a finite binding energy. Frequently this difficulty is avoided by cutting off the potentials arbitrarily at a certain minimum distance. The procedure is obviously unsatisfactory, especially in view of the fact that the cut-off distance turns out to be of the order of magnitude of the range of the nuclear forces. These failures of the theory are not surprising. It has already been noticed in Section 11 that the development parameter of the perturbation theory adopted in treating the Yukawa problem is not at all small. This means that there is no justification for the assumption that the second approximation calculation should give a reliable result. It is not clear, however, whether a mathematically consistent treatment of the Yukawa problem is at all possible, and no such treatment is known.

We conclude this section by mentioning a simplified procedure for arriving at the general form of the Yukawa potential. This procedure will be illustrated in the case of the interaction between two nucleons due to a neutral pion field. This field will be treated as a non-quantized classical field obeying the Klein-Gordon equation (10). The assumption will be made that the nucleons act as sources of the pion field in the same way that electric charges are the sources of the electrostatic field. The force between two nucleons is due to the action of the field produced by one nucleon on the other. This is the exact analogue of the description of the electrostatic force between two charges as due to the action of the electric field produced by one charge on the other. The spherically symmetrical solution of the Klein-Gordon equation with a point source of strength e_2 is found to be

$$\varphi = \frac{e_2}{4\pi} \frac{1}{r} e^{-kr}. \tag{100}$$

We regard φ as the analogue of a potential which acts on the "charge" of the second nucleon causing an interaction energy

$$\frac{e_2^2}{4\pi r} e^{-kr}. \tag{101}$$

This mutual energy has the form of the Yukawa potential since k is related to the pion mass by (14).

18. PRODUCTION OF PIONS IN NUCLEON COLLISIONS

The qualitative prediction that mesons should be produced in the collision of fast nucleons has been one of the greatest achievements of the Yukawa theory. Indeed, not only does the theory predict qualitatively this phenomenon that has been since observed both in cosmic radiation events and in synchrocyclotron bombardments but it also predicts correctly the order of magnitude of the cross section to be expected.

In outlining the theory of this phenomenon no distinction will be made at first between neutrons and protons. Nucleons will be indicated by the general symbol N. Also no distinction will be made between charged or uncharged pions which will be indicated by Π. The process that we are discussing corresponds to the reaction

$$N_1 + N_2 \rightarrow N_3 + N_4 + \Pi. \tag{102}$$

From the point of view of the perturbation theory this is a third approximation process in spite of the fact that only one pion ultimately is emitted. Indeed, a first step in which the pion is emitted by one of the nucleons, for example N_1, according to the scheme

$$N_1 + N_2 \rightarrow N_1' + N_2 + \Pi, \tag{103}$$

is evidently incompatible with the simultaneous conservation of energy and momentum, since in the rest system of the nucleon N_1 there is no energy available for the formation of the pion, and the second nucleon N_2 does not contribute to the process. Formula (103) can be therefore only one of several steps needed to achieve the reaction (102). The other steps involve an interaction between N_1' and N_2 such as could be obtained for example by emission of another pion by one of the two nucleons and reabsorption by the other. So, in all there are three steps.

A detailed analysis of the case shows that the transition (102) may take place through a variety of intermediate steps. All their contributions should be added in order to form the effective matrix element corresponding to the transition. Since, however, we are trying to compute merely the order of magnitude of the probability of pion formation, it will suffice to calculate the contribution of only one set of intermediate states. As such, we choose the sequence of steps outlined above. The calculation may be further simplified. It has been shown in the previous section that emission of a pion by a nucleon and reabsorption by the other nucleon is equivalent in terms of the Yukawa theory to a force acting between the two nucleons. The corresponding matrix element is given by formula (93). The transition (102) will be achieved therefore in two steps, a first step (103) and a second step

$$N_1' + N_2 + \Pi \rightarrow N_3 + N_4 + \Pi \tag{104}$$

which summarizes the effect of the absorption and emission of a virtual pion.

The matrix element corresponding to the transition (103) is given by (42). The step (104) contributes the matrix element (93). In a computation of order of magnitude we will simplify both (42) and (93) by substituting μc^2 for w. The effective matrix element for the process in question will be the product of (42) and (93) divided by the energy difference between initial and intermediate states. This again will be of the order of magnitude μc^2.

In conclusion, the order of magnitude of the effective matrix element for the process (102) is

$$f/\Omega^{3/2} \quad \text{where} \quad f = \frac{e_2^2\,\hbar^3}{\sqrt{2}\,\mu^{7/2}c^4}. \tag{105}$$

Numerically one finds by substitution $f \approx 10^{-61}$.

The computation of the rate of transition can now be carried out in the center-of-mass system for two nucleons colliding with relative velocity v according to the usual pattern of Section 10. The partial cross section σ_{dp} is calculated for the production of a pion of momentum between p and $p + dp$. The rate of the transition $v\sigma_{dp}/\Omega$ can be calculated according to (55). Since three particles are emitted the factor dN/dW of (55) should be computed as in Section 15, with the difference, however, that in that case the three particles were extreme relativistic. In the present case it will be assumed instead that the pion and the two nucleons N_3 and N_4

are classical. The momentum p of the pion and the momenta p_3 and p_4 of the nucleons will be related by the energy-conservation equation

$$\frac{1}{2M}(p_3^2 + p_4^2) + \frac{1}{2\mu}p^2 = T \tag{106}$$

in which T is the amount of energy in the center-of-mass system above the threshold for pion production. Because of the momentum conservation the vector sum $p + p_3 + p_4$ vanishes and the three vectors form a triangle like the three vectors in Figure 2. Since, however, in the present case (106) holds instead of (80), one finds that the locus of the vertex A of the triangle is not an ellipsoid but is a sphere of radius,

$$\left\{ MT - \left(\frac{1}{4} + \frac{M}{2\mu}\right)p^2 \right\}^{1/2}.$$

As in Section 15, dN/dW is the derivative with respect to the energy T of the product of the volume of this sphere by the factor $\Omega/(2\pi\hbar)^3$, namely

$$\frac{dN}{dW} = \frac{\Omega M}{4\pi^2 \hbar^3} \left\{ MT - \left(\frac{1}{4} + \frac{M}{2\mu}\right)p^2 \right\}^{1/2}. \tag{107}$$

One can now compute the rate of transition according to (55). Since the number of final pion states is $\Omega p^2 dp/2\pi^2\hbar^3$, the rate must be multiplied by this factor. This rate can also be expressed in the form $v\sigma_{dp}/\Omega$. One finds then

$$\sigma_{dp} = \frac{f^2 M}{4\pi^3 \hbar^7}\frac{1}{v} \left\{ MT - \left(\frac{1}{4} + \frac{M}{2\mu}\right)p^2 \right\}^{1/2} p^2\, dp. \tag{108}$$

The total cross section for pion production is obtained by integrating (107). The result is

$$\sigma = \frac{f^2 M^3}{64\pi^2 \hbar^7}\frac{1}{v}\frac{T^2}{\left(\dfrac{1}{4} + \dfrac{M}{2\mu}\right)^{3/2}}. \tag{109}$$

For bombarding nucleons of 345 Mev one finds from this formula in agreement with the Berkeley experiments a cross section of the order of 10^{-28} cm^2.

Formula (108) also gives the shape of the pion spectrum plotted in Figure 4. The curve is a circle.

In the production of pions by proton-proton collisions, however, the observed shape of the spectrum is quite different. This is because the wave functions of the two nucleons produced after the process are distorted by action of the nuclear forces. The effect of this complication will be discussed in the next section.

19. EFFECT OF THE NUCLEON BOND

In discussing the production of pions in a collision of two nucleons it has been assumed that the states of the nucleons and of the pions before and after the collision are represented by plane waves. This assumption is allowable when the relative

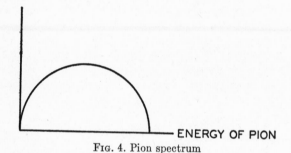

Fig. 4. Pion spectrum

energy of the particles is very high so that the distortion of the wave functions due to the nuclear forces may be neglected. Otherwise it may lead to serious error.

In particular, in the case of the production of positive pions in a collision of two protons according to the reaction

$$P_1 + P_2 \rightarrow P_3 + N_4 + \Pi^+ \tag{110}$$

the forces between the proton and the neutron produced after the reaction, as pointed out by Chew, may be so important that the two particles may escape bound as a deuteron. Also the shape of the spectrum of the pions is quite different from the one represented in Figure 4. In the experiments carried out at Berkeley on this reaction by Peterson and by Cartwright, Richman, Whitehead and Wilcox, the energy of the bombarding protons was 345 Mev. The amount of energy available in the center-of-gravity system is about one-half this amount—precisely, 165 Mev—including relativity corrections. Since the relative energy of the two protons P_1 and P_2 is so great, no large error is introduced by neglecting the distortion of their wave functions.

However, after the reaction has taken place and a pion has been produced with a mass energy of 140 Mev the residual energy is only 25 Mev, part of which is taken away as kinetic energy of the pion. Therefore the relative energy of the two nucleons P_3 and N_4 is small and the effect of the distortion of their wave functions becomes conspicuous. The effect will be most pronounced when the pion escapes with energy near the maximum possible since then the relative velocity of P_3 and N_4 will be quite small.

The effect will be estimated by including in the discussion only the force between P_3 and N_4. In the previous section we have calculated the order of magnitude of the matrix element for a process like (102) or (110). The value (105) that was found for this matrix element must now be modified as follows. If the two nucleons P_3 and N_4 could be represented by plane waves as was assumed in the previous section, their combined wave function would be

$$\frac{1}{\sqrt{\Omega}} e^{(i/\hbar) p_3 \cdot r_3} \times \frac{1}{\sqrt{\Omega}} e^{(i/\hbar) p_4 \cdot r_4}$$

$$= \frac{1}{\sqrt{\Omega}} e^{(i/\hbar)(p_3 + p_4) \cdot r_D} \times \frac{1}{\sqrt{\Omega}} e^{(i/\hbar)\{(p_3 - p_4)/2\} \cdot (r_3 - r_4)} \tag{111}$$

where $r_D = (r_3 + r_4)/2$ is the position vector of the center of gravity of the two nucleons. The eigenfunction (111) has been rearranged as the product of a factor representing the wave function of the center of gravity of the two nucleons and a factor representing the relative motion. It is this last factor only that will be distorted by the attraction between P_3 and N_4. The wave function for the relative motion will no longer be a plane wave but will be represented by a more complicated function, $D(r_3 - r_4)$. For example, if the two particles escape combined as a deuteron, the function D would be the deuteron wave function. If they are not bound, the wave function of a dissociated state of the deuteron should be used.

Since the two particles P_3 and N_4 are created at quite close distance, the matrix element for the transition (110) will be proportional to the value of the wave function D taken for $r_3 = r_4$, that is, to $D(0)$. If the binding forces were neglected D would reduce to the last factor of (111) and therefore $D(0)$ would equal $1/\sqrt{\bar{\Omega}}$. Consequently, the matrix element (105) wll be corrected by a factor equal to the ratio

$$D(0)/(1/\sqrt{\bar{\Omega}}) = \sqrt{\bar{\Omega}}D(0) \qquad (112)$$

and become

$$fD(0)/\Omega. \qquad (113)$$

The constant f can be expressed in terms of e_2 according to (105):

$$f = \frac{e_2^3 \hbar^3}{\sqrt{2}\,\mu^{7/2} c^4}. \qquad (114)$$

Since, however, this formula was derived by a somewhat questionable application of the perturbation theory, (114) will give at best the order of magnitude of f. On the other hand, as long as the pion is formed in states of rather low energy we can expect f to be fairly constant and use (113) in order to determine the shape of the pion spectrum and the probabilities that the proton and the neutron are emitted either free or bound as a deuteron. The value of f that fits the experimental results will then be compared with its value (114) obtained from the perturbation calculation of the last section.

We first calculate the cross section for a process in which P_3 and N_4 are bound as a deuteron. The result of the reaction (110) will be two particles, a pion and a deuteron, escaping with opposite momentum. In the matrix element (113), $D(0)$ will be the value of the deuteron wave function for zero distance. Its numerical value can be obtained by computing the deuteron wave function on the assumption that the attraction between proton and neutron is due to a square-well potential with radius 2.82×10^{-13} cm and depth 21 Mev. By normalizing the wave function properly one finds that its value at the origin is 3.2×10^{18} cm$^{-3/2}$. The rate of the reaction will be calculated with the usual formulas (55) and (60) whereby the sum of the speeds of the pion and of the deuteron will appear in the denominator. The rate of the reaction may also be expressed in the form $\sigma_D v_{12}/\Omega$ where σ_D is the cross section for the process in which a deuteron is formed and v_{12} is the relative velocity of the two colliding nucleons. One finds

$$\sigma_D = \frac{f^2 D^2(0) p^2}{\pi \hbar^4 (\,|\,v_D\,| + |\,v_\pi\,|\,)v_{12}} = \frac{f^2 D^2(0) M_R p}{\pi \hbar^4 v_{12}}. \qquad (115)$$

M_R is the reduced mass of the deuteron and the pion: $M_R = 2M\mu/(2M + \mu)$. The momentum p is calculated from the conservation of energy $p^2/2M_R = T + B$ where T is the kinetic energy of P_3, N_4 and Π available in the center-of-gravity system exclusive of the rest energies, and B is the binding energy of the deuteron equal to 2.2 Mev. The numerical results will be discussed at the end of this section.

Formula (115) has been derived for P_3 and N_4 bound in a deuteron state. The same formula can be applied also if the two particles are not bound but escape from each other with momentum q. The only difference is that the function D will now represent the wave function of the deuteron in a dissociated state. This also can be computed by standard procedures assuming a square-well potential of radius r_0 and depth U. $D(0)$ is different from zero only for s-states. The value of $D^2(0)$ calculated for an s-state should then be multiplied by the number of s-states of relative momentum between q and $q + dq$. The calculation yields

$$D^2_{\text{cont}}(0) = \frac{q^2\,dq}{2\pi^2\hbar^3}\,\frac{q^2 + MU}{q^2 + MU\cos^2\left(\sqrt{q^2 + MU}\,r_0/\hbar\right)}. \qquad (116)$$

Substituting in (115) one obtains the cross section σ_{dq} for a process in which P_3 and N_4 escape with a relative momentum between q and $q + dq$

$$\sigma_{dq} = \frac{f^2 M_R\,p q^2\,dq}{2\pi^3\hbar^7 v_{12}}\,\frac{q^2 + MU}{q^2 + MU\cos^2\left(\sqrt{q^2 + MU}\,r_0/\hbar\right)}. \qquad (117)$$

q and p are related by the conservation of energy,

$$\frac{q^2}{M} + \frac{p^2}{2M_R} = T, \qquad (118)$$

where T is the kinetic energy of the products of the reaction. With the help of this equation the formula (117) may be transformed into

$$\sigma_{dp} = \frac{f^2 M q p^2\,dp}{4\pi^3\hbar^7 v_{12}}\,\frac{q^2 + MU}{q^2 + MU\cos^2\left(\sqrt{q^2 + MU}\,r_0/\hbar\right)} \qquad (119)$$

where σ_{dp} represents the cross section for a process in which the pion is emitted with a momentum between p and $p + dp$ in the center-of-gravity system. This formula is more convenient if one wants the shape of the spectrum of the emitted pions. For $U = 0$ it becomes identical to (108).

So far we have disregarded the spin of the two nucleons produced. Actually, P_3 and N_4 may escape with parallel spins in which case they may be either bound as a deuteron or dissociated. Or they may have anti-parallel spins, in which case no bound state exists. Consequently the spectrum of the pions emitted will be a

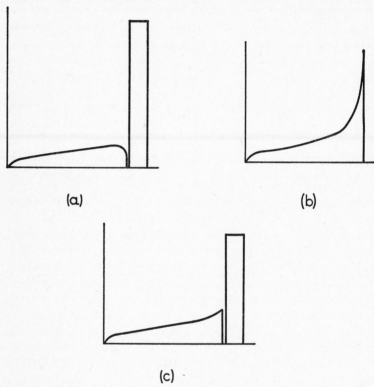

(a)   (b)

(c)

FIG. 5. Shapes of the pion spectrum in the center-of-gravity frame. The line at the high-energy end of the curves (a) and (c) has been given a finite thickness in order to show its intensity relative to the continuous spectrum.

superposition of two spectra corresponding to these two cases. Since the interactions are presumably spin dependent, we cannot be sure that the constant f has the same value for the two cases. However, if we make this assumption, the two spectra will be superimposed in the ratio of 75% for the spectrum corresponding to parallel spins and 25% for the other. In Figure 5 the curves (a), (b), and (c) correspond to the spectrum expected for spins parallel or anti-parallel, or for the 75%-25% mixture, and for bombarding energy of 345 Mev.

The resolution of the experimental results is not yet adequate to distinguish between the three possibilities represented in Figure 5. In particular the single line appearing on the high-energy side in cases (a) and (c) cannot be resolved from the continuous spectrum. In any case there is a very obvious indication of what appears as a sharp maximum at the high-energy side of the spectrum. This maximum could be consistent with any of the three shapes indicated. Assuming for example the 75%-25% mixture of curve (c), one can further derive from the comparison of the experimental and theoretical cross section the value of the constant f. One finds $f = 6.5 \times 10^{-62}$ ergs $\times$ cm$^{9/2}$. This value may be compared with the value calculated according to (114) which is 13×10^{-62}. This may be regarded as a rather good agreement since the theory is too coarse to be trusted to better than the order of magnitude.

The forces acting between the two outgoing nucleons not only change the shape of the pion spectrum from that of Figure 4 to those of Figure 5, they also cause a change of the total cross section, by about a factor 3.4, for bombarding energy of 345 Mev.

When positive pions are formed by bombarding a carbon target with protons one might expect to observe a cross section for formation of positive pions at least six times as large as the cross section observed in the bombardment of hydrogen with protons, since a carbon nucleus contains six protons. The observed value is actually appreciably less. The fact that bombarding a free proton gives a higher yield than bombarding the protons bound in the carbon nucleus probably is due to the resonance of a free proton and neutron at low energies. This resonance is no longer there when the nucleons are not free.

In a collision of two high-energy protons not only positive pions can be formed according to (109) but also neutral pions according to the reaction

$$P_1 + P_2 \to P_3 + P_4 + \Pi^0. \tag{120}$$

Experimentally the yield of this reaction is found to be less than 10 or 20 per cent of that of reaction (109). The reason for this fact is somewhat obscure. Since P_3 and P_4 are two identical particles

they can be produced only in a singlet state, so that the spectrum
of the neutral pions emitted could be somewhat similar to spectrum
(b) in Figure 5. The maximum would be somewhat less pronounced
on account of the Coulomb repulsion of the two protons, but no
very large difference is expected. Perhaps the failure to observe
neutral pions in this case may be due to the fact that the constant
f for the singlet emission is appreciably smaller than for the
triplet emission.[2]

20. PRODUCTION OF PIONS BY GAMMA RAYS

Experiments conducted at Berkeley by McMillan,
Steinberger, Panofsky, and others with gamma rays produced in
the synchrotron with a maximum energy of 335 Mev have proved
that pions both charged and neutral are produced by high-energy
photons impinging on nucleons. The theory of the photo-production
of pions will be sketched here.

We take as an example the production of negative pions by
photons colliding with a neutron according to the reaction

$$\gamma + N \rightarrow P + \Pi^-. \tag{121}$$

Since a pion is produced and a photon disappears, this reaction
will be a second approximation perturbation process. For example,
we may consider the following two steps. A first step

$$\gamma + N \rightarrow \gamma + P + \Pi_1^- \tag{122}$$

in which the neutron converts into a proton and a pion according
to the Yukawa process (38b). As a result of this step in which the
photon is not affected, an intermediate state will be reached with
energy larger than the energy of the initial state by about μc^2.
The order of magnitude of the matrix element is given by (42)
in which we substitute for w_s its order of magnitude μc^2. The
matrix element of the first step is thus

$$\frac{e_2 \hbar}{\sqrt{2\Omega\mu}}. \tag{123}$$

---

2. Recent results from Berkeley seem to indicate that the angular distribution of the pions produced in the reactions (109) is not isotropic in the
center-of-mass system but may perhaps be better represented by a $\cos^2 \vartheta$
distribution. This would indicate a strong interaction with p-waves and
corresponding changes should be introduced in the discussion of the last
two sections.

A second step

$$\gamma + P + \Pi_1^- \rightarrow P + \Pi^-$$ (124)

is required in order to reach the final state of (121). This is caused by the electromagnetic interaction between the photon and the pion. The photon is absorbed by the pion, which thereby changes its momentum to the final value. The matrix element for this process is of the order (35), since the velocity of the pion is of the order of c. The matrix element therefore has the order of magnitude

$$\frac{ec\sqrt{2\pi\hbar}}{\sqrt{\Omega\omega}}.$$ (125)

The matrix element for the composite transition (121) is obtained as usual from the perturbation formula (61) and is the product of (123) and (125) divided by the energy difference between intermediate and initial states which, as we have seen, is of the order of magnitude μc^2. Again we expect to obtain the correct order of magnitude by considering only one of several possible intermediate states. The result is therefore

$$\mathcal{H}'_{n0} = \frac{\sqrt{\pi}\, ee_2\, \hbar^{3/2}}{\Omega c \mu^{3/2}\, \omega^{1/2}}.$$ (126)

The probability of transition from initial to final state is calculated as usual with (55). Since the photon travels with the velocity of light the transition takes place at a rate $\sigma c/\Omega$. Using (55) and (58) one obtains

$$\frac{\sigma c}{\Omega} = \frac{2\pi}{\hbar}\, |\,\mathcal{H}'_{n0}\,|^2\, \frac{\Omega p^2}{2\pi^2 \hbar^3 v}$$ (127)

in which p and v are the momentum of the pion and the relative velocity of the pion and the proton. The small proton velocity being neglected, (127) yields the final result

$$\sigma = \frac{e^2 e_2^2}{\hbar c^3 \mu^3}\, \frac{p^2}{\omega v} \approx \frac{\sqrt{2}\, e^2 e_2^2}{c^3\, \mu^{3/2}}\, \frac{\sqrt{\hbar\omega - \mu c^2}}{\hbar\omega}.$$ (128)

Substituting numerical values in this formula we find for photons of 335 Mev a cross section of the order of 3×10^{-28} cm^2.

In the previous outline of the theory of this process some of the possible intermediate steps have been omitted. For example, we have assumed that in the first step (122) the proton is created directly in the final state and the pion has a momentum different from the final one. In the second step we assume that the pion absorbs the photon, changing its momentum to the final value. Since, however, the photon may react not only on the pion but also on the proton, the contribution of an intermediate state $\gamma + P_1 + \Pi^-$ should be included. This is a state in which the pion is directly formed by the Yukawa process into its final state while the proton is formed in an intermediate state P_1 which changes to the final state after absorption of the photon.

The contribution of a transition obtained in this second manner should be included in computing the effective matrix element. It should be noticed in particular that the situation is somewhat dependent on whether positive pions are produced by the action of gamma rays on neutrons or whether negative pions are produced by the action of gamma rays on protons. This difference has been analyzed by Goldberger who has found that it explains the observed difference in the cross sections for production of positive and negative pions.

A final remark on the production of neutral pions: Since these particles presumably do not interact strongly with the electromagnetic field, the intermediate state (122) cannot be used and the photon will be able to interact only with the nucleon. This would seem to indicate that the cross section for production of neutral pions by action of photons on hydrogen should be appreciably less than the similar cross section for the production of charged pions. This expectation seems to be contradicted by the experimental results of the Berkeley group. Their results indicate that the cross sections for the two processes are about equal.

21. CAPTURE OF NEGATIVE PIONS BY PROTONS

Panofsky, Aamodt, and York have investigated the phenomena that take place when a negative pion comes to rest in hydrogen. They found that gamma rays are emitted with a spectral distribution which consists of two groups: one at about 60 or 70 Mev and one at about 120 Mev. The following theoretical interpretation has been suggested by Marshak. A negative pion that loses its kinetic energy by ionization and comes to rest in a hydrogen atmosphere is promptly attracted to one of the protons in the vicinity and has an appreciable probability of being captured in the K-orbit near the proton before disintegrating into a muon and a neutrino according to the process discussed in Section 13. As soon as the pion comes close to the proton different and more rapid processes of destruction of the pion become possible. The following three possibilities have been considered: A transformation of the proton and the negative pion into a neutron and a gamma ray:

$$P + \Pi^- \rightarrow N + \gamma. \tag{129}$$

If this reaction takes place the mass energy of the pion of about 140 Mev will be partitioned between the photon and the neutron which escape with equal and opposite momenta. Conservation of energy and momentum lead one to expect the emission of a photon of about 131 Mev and a neutron of about 9 Mev. This process therefore could account for the high energy component of the observed gamma spectrum.

The lower energy component, however has a frequency which suggests that the bulk of the energy of the pion is converted into two photons instead of one. Two processes have been considered. One,

$$P + \Pi^- \rightarrow N + 2\gamma, \tag{130}$$

is the direct transformation of the original pion and proton into a neutron and two gamma rays. The other one assumes that the pion and proton first change into a neutron and a neutral pion. Shortly afterward the neutral pion disintegrates into two photons according to a process that will be discussed in detail in Section 24. The reaction therefore would be

$$P + \Pi^- \rightarrow N + \Pi^0 \tag{131}$$

and very shortly afterward

$$\Pi^0 \rightarrow 2\gamma. \tag{132}$$

Since the last process is believed to take place in a time shorter than 10^{-14} seconds, the whole phenomenon will be controlled by the rate of the reaction (131).

Reaction (131) will be discussed first. It is energetically possible only provided that the mass energy of the neutron and of the neutral pion is less than the energy of the bound proton and negative pion. If we assume this to be the case and call ϵ the excess energy available, the process can be achieved in two steps, as follows:

$$P + \Pi^- \to N_1 \to N + \Pi^0. \tag{133}$$

These two steps correspond to the elementary processes (38b) and (38d) of the Yukawa interaction. The matrix elements of both are given as order of magnitude by (42) in which w in the denominator will be replaced by μc^2 since both the charged and the neutral pion have very low kinetic energy. Therefore the order of magnitude of the two matrix elements is

$$\frac{e_2 \hbar}{\sqrt{2\Omega\mu}}.$$

The intermediate state N_1 of (133) consists of a nucleon at rest. Its energy is therefore lower than the energy of the initial state by about μc^2. We can now compute the apparent matrix element for the transition from the initial to the final state of (133). Using (61) one finds that the contribution of (133) to its value is

$$\frac{1}{\mu c^2} \left(\frac{e_2 \hbar}{\sqrt{2\Omega\mu}} \right)^2 = \frac{e_2^2 \hbar^2}{2\Omega\mu^2 c^2}. \tag{134}$$

The intermediate state N_1 of (133) is not the only possibility. Another possible intermediate state is $\Pi^- + P_1 + \Pi^0$. In other words, instead of there being absorption of the negative pion first and then emission of Π^0, the order of these two steps may be reversed.

One of the two possible intermediate states, N_1, is energetically below the initial state by about μc^2, and the other, $\Pi^- + P_1 + \Pi^0$, is above by the same amount. If the two contributions to the effective matrix element were added according to (61) they might almost cancel. The transition (131) would then be very improbable. Since this apparently is not the case one must conclude that there is no important cancellation. This could be due, for example, to the fact that the coupling constants of the neutral pions to the protons and to the neutrons are equal but of opposite sign. The contributions of the two intermediate states would then be added instead of subtracted. In our estimate of the order of magnitude the contribution of only one intermediate state has been included and (134) has been used.

The probability of transition for the process (131) is now calculated according to the usual pattern with (55) and (60). One finds

$$\text{rate of (131)} = \frac{1}{\Omega} \frac{e_2^4 p^2}{4\pi\mu^4 c^4 (v_N + v_\Pi)}$$
$$= \frac{e_2^4 M^{3/2} \sqrt{\epsilon}}{2\sqrt{2}\pi c^4 \mu^{5/2}(M + \mu)^{3/2}} \frac{1}{\Omega} \qquad (135)$$

where p is the momentum of the neutral pion which has been computed from the conservation of energy:

$$\left(\frac{1}{2M} + \frac{1}{2\mu}\right) p^2 = \epsilon. \qquad (136)$$

The volume factor Ω appears in (135) because the rate of the reaction (131) is proportional to the concentration of protons to which the negative pion is exposed. If initially the two particles are bound in a K-like orbit, Ω should be replaced as in Section 16 by the effective volume of this orbit,

$$\Omega_k = \pi(a/276)^3, \qquad (137)$$

where a is the Bohr radius and 276 is the ratio of the masses of the pion and of the electron. One finds

$$\text{rate of (131)} = \frac{1}{2^{3/2}\pi^2} \left(\frac{276}{a}\right)^3 \frac{M^{3/2} e_2^4 \sqrt{\epsilon}}{c^4 \mu^{5/2}(M + \mu)^{3/2}} \qquad (138)$$
$$= 2 \times 10^{15} \sqrt{\epsilon_{\text{Mev}}}.$$

Shortly after being emitted the neutral pion disintegrates into two photons. If the neutral pion had zero velocity a monochromatic radiation with photons of one-half the mass energy of the neutral pion would be emitted. If, however, the neutral pion escapes with some amount of kinetic energy the spectral distribution will be broadened by Doppler effect. The width of the gamma-ray peak at about 70 Mev therefore gives a very sensitive means of comparing the masses of the neutral and the charged pion. According to Panofsky and his co-workers the present experimental evidence indicates that the difference in mass of the negative and the neutral pion is about 5 Mev. The rate of (131) is therefore about 4×10^{15} per second.

We will not discuss in detail the process (130). Its rate is appreciably slower than that of (131). Also the spectrum of the photons emitted would be difficult to reconcile with the sharp peak observed experimentally near 70 Mev.

The rate of the reaction (129) that may be responsible for the 130 Mev gamma rays also can be computed as a second approximation process according to the scheme

$$P + \Pi^- \rightarrow \Pi_1^- + \gamma + P \rightarrow N + \gamma. \tag{139}$$

The intermediate state is reached by a first process in which the negative pion emits the gamma ray. The matrix element for this process has the order of magnitude (35). With $\omega \approx \mu c^2/\hbar$ this gives

$$\sqrt{2\pi}e\hbar/\sqrt{\Omega\mu}. \tag{140}$$

The second step of (139) corresponds to the Yukawa process (38b) and its matrix element, as before, is of the order of magnitude

$$e_2\hbar/\sqrt{2\Omega\mu}. \tag{141}$$

Again the energy of the intermediate state differs from that of the initial state by about μc^2. One can therefore calculate the rate of the process by the usual procedure of Section 10 and one finds

$$\begin{aligned}
\text{rate of (139)} &= \frac{1}{\Omega_K} \frac{e^2 e_2^2}{\mu^2 c^3} \\
&= \frac{1}{\pi} \left(\frac{276}{a}\right)^3 \frac{e^2 e_2^2}{\mu^2 c^3} \approx 6 \times 10^{14} \text{ sec}^{-1}.
\end{aligned} \tag{142}$$

It appears therefore that the rate of the process (139) should be comparable to but probably somewhat slower than that of the process (131). Experimental results seem to indicate that the two processes have about the same rate.

22. SCATTERING OF PIONS BY NUCLEONS

According to the Yukawa theory one expects that pions will be scattered when colliding against nucleons. When both pions and nucleons are electrically charged the Coulomb scattering will be superimposed on the scattering due to the Yukawa interaction. At high energies, however, the Yukawa scattering should be dominant. As a first example the Yukawa scattering of a positive pion by a neutron will be discussed.

This process is described in the perturbation theory as one of second approximation since the pion in the original momentum state must be destroyed and the scattered pion in the new momentum state must be created. The two steps of the process could be, for example:

$$N_1 + \Pi_1^+ \to P \to N_2 + \Pi_2^+ . \qquad (143)$$

Both steps are determined by the Yukawa interaction (38a). In the center-of-gravity system the proton in the intermediate state[3] will have zero momentum.

Assuming for simplicity that the kinetic energy of the neutron is negligible, one finds that the energy of the intermediate state is lower than that of the initial state by an amount equal to the energy w of the pion. Both matrix elements for the two steps are given by (42). The apparent matrix element for the transition from the initial to the final state of (143) is therefore, according to (61),

$$\mathcal{H}'_{21} = \frac{1}{w} \left(\frac{e_2 \hbar c}{\sqrt{2\Omega w}} \right)^2 = \frac{e_2^2 \hbar^2 c^2}{2\Omega w^2}. \qquad (144)$$

From this matrix element one computes the scattering cross section by the usual procedure using (55) and (56). The result is

$$\sigma = \frac{e_2^4 c^4 p^2}{4\pi w^4 v^2} = \frac{1}{4\pi} \left(\frac{e_2^2}{w} \right)^2 \qquad (145)$$

where p, v, w, are momentum, velocity, and total energy of the scattered pion in the center-of-gravity system. In deriving the last expression the relationship $c^2 p = vw$ has been used. The cross section (145) has its maximum value for pions of low velocity and is inversely proportional to the energy of the pion (including rest energy). Adopting our usual value $e_2 = 10^{-8}$, one finds that

---

3. In this case also another intermediate state, $\bar{P}_3 + \Pi_2^+ + \Pi_1^+$, might contribute. $\bar{P}_3$ is an anti-proton. For this state, however, the energy denominator of (61) would be very large. Hence its contribution would be small.

the cross section (145) at low energies is 1.6×10^{-26} cm^2. The energy dependence indicated by formula (145) is not reliable since it has been deduced from the simplified form (41) of the Yukawa interaction. Other forms of interaction could give a quite different and even an opposite dependence on the energy. It should further be noted that the neutron-pion scattering could be described as due to a potential between the neutron and the pion, having the matrix element (144). This would be the analogue of the Yukawa theory of nuclear forces. The two steps (143) would be interpreted as equivalent to a force acting between the neutron and the pion.

The four Yukawa processes (38) allow similar scattering processes for the following additional cases besides (143):

$$P_1 + \Pi_1^- \rightarrow P_2 + \Pi_2^-$$
$$P_1 + \Pi_1^0 \rightarrow P_2 + \Pi_2^0 \qquad (146)$$
$$N_1 + \Pi_1^0 \rightarrow N_2 + \Pi_2^0 .$$

In the last two cases, however, the cross section could be smaller due to a possible destructive interference of two intermediate states such as P_3 and $P_3 + \Pi_1^0 + \Pi_2^0$. Additional processes which are possible according to (38) and which lead to a cross section of the order (145) are

$$\Pi_1^+ + N_1 \rightarrow P_3 \rightarrow \Pi_2^0 + P_2$$
$$\Pi_1^- + P_1 \rightarrow N_3 \rightarrow \Pi_2^0 + N_2 \qquad (147)$$

and their inverse processes. In these cases the scattering is accompanied by an exchange of the charge between the pion and the nucleon. For example, the first of the processes (147) would be obtained by a successive application of the interactions (38a) and (38c).

The reaction (131) discussed in the previous section is quite similar to the second of the reactions (147). Probably the mass of the neutral pion is slightly smaller than that of the charged pion. Hence the kinetic energy of the neutral pion and the neutron will be slightly larger than that of the negative pion and the proton. The reaction can therefore take place even when the proton and the negative pion are bound in a K-orbit as has been seen in the previous section.

The Yukawa reactions (38) do not allow a quite similar process for the scattering of a positive pion by a proton or for that of a negative pion by a neutron. However, these processes can also take place in the same approximation and with cross sections of the same order of magnitude, by two steps as follows:

$$\Pi_1^+ + P_1 \rightarrow \Pi_1^+ + N_3 + \Pi_2^+ = \Pi_2^+ + P_2 \qquad (148)$$

in which the interaction (38a) or (38b) operates twice in succession.

23. THE ANTI-NUCLEON. ANNIHILATION

All the current theories of electrically charged particles have a symmetry property according to which for each particle a counterpart with the opposite charge and otherwise similar properties exists. This is true in particular of the Dirac electron theory which was established before the discovery of the positron. In most discussions about nucleons these particles are supposed to obey a Dirac-like equation. If this assumption is correct, negative protons must exist and also anti-neutrons. The anti-proton, here indicated by $\bar{P}$, has the mass of the proton and has negative charge and magnetic moment equal and opposite to that of the proton. The anti-neutron, indicated by $\bar{N}$, has the mass of the neutron, no charge, and magnetic moment equal and opposite to that of the ordinary neutron.

Since no experimental evidence has been found for the existence of these two particles we cannot be too sure that they really exist. It is interesting, nevertheless, to speculate as to what their properties are likely to be. In this discussion the somewhat similar case of the behavior of electrons and positrons may be taken as a guide.

One might expect in particular that when an anti-nucleon comes in contact with an ordinary nucleon the two particles may annihilate each other, releasing their mass energy $2Mc^2$. This annihilation process may occur through the interaction of the nucleons with the electromagnetic field, in which case the result of the reaction will be the emission of two photons with equal and opposite momentum as in the case of the electron-positron annihilation. The annihilation may also be caused by the coupling of the nucleons to the pion field, in which case the product of the reaction will be two pions with equal and opposite momenta. Since nucleons are more strongly coupled to pions than they are to the electromagnetic field, this last process will be more probable than the first. In spite of this fact, the annihilation with two-photon emission will be treated first because its results will be needed for the calculations in the next section.

We propose, therefore, to calculate the rate of annihilation of a proton and an anti-proton of very low kinetic energy with emission of two photons. The conservation of momentum and energy requires that the two photons should escape in opposite directions and have energy $w = Mc^2$ each. Since two photons must be emitted, the simplest process is one of second approximation in which one proceeds from the initial to the final state through an intermediate state in which only one photon has been emitted according to the reaction

$$P_1 + \bar{P}_1 \rightarrow P_2 + \bar{P}_1 + \gamma_1 \rightarrow \gamma_1 + \gamma_2 \,. \tag{149}$$

In an exact calculation one should actually take into account several possible intermediate states. For example, the photon emitted in the intermediate state could have been emitted by the anti-proton. Also a number of different spin possibilities should be considered. In evaluating orders of magnitude, however, it will suffice to consider one intermediate state only. The intermediate state will differ in energy from the initial state by amounts of the order of Mc^2. For example, if the photon in the intermediate state is emitted by the proton, this particle that was initially at rest acquires in the intermediate state the recoil momentum Mc and the kinetic energy $(\sqrt{2} - 1)Mc^2$. To this must be added the energy Mc^2 of the emitted photon so that the energy of the intermediate state rises above the energy of the initial state by the amount $\sqrt{2}Mc^2$. The same would be true if the photon had been emitted by the anti-proton.

The order of magnitude of the matrix elements corresponding to the two steps of (149) can be estimated from (35) since the proton P_2 in the intermediate state has a velocity comparable to that of light.

With $\omega_s = Mc^2/\hbar$ the result is

$$\frac{\sqrt{2\pi}\, e\hbar}{\sqrt{\Omega M}} \,. \tag{150}$$

We now apply (61) in order to calculate the apparent matrix element $\mathcal{H}'_{no}$. This is the product of two factors like (150) corresponding to the matrix elements of the two steps of (149) divided by the difference in energy between initial and intermediate states which is $\sqrt{2}Mc^2$. The result is

$$\mathcal{K}'_{no} = \frac{1}{\sqrt{2}Mc^2}\left(\frac{\sqrt{2\pi}\,eh}{\sqrt{\Omega M}}\right)^2 = \frac{\sqrt{2}\,\pi e^2 \hbar^2}{\Omega M^2 c^2}. \qquad (151)$$

The calculation of the probability of transition proceeds now according to the usual rules. The density factor dN/dW can be estimated with (60). Both v_1 and v_2 are equal to c and p equals Mc. One finds

$$\text{rate of annihilation} = \frac{\pi e^4}{M^2 c^3}\frac{1}{\Omega}. \qquad (152)$$

An accidental cancellation of the numerical factors neglected here and there makes this formula identical with the one that would be obtained with a more elaborate theory.

It is not surprising that the normalization volume Ω does not disappear from the final formula since the anti-proton is annihilated by one proton present in the volume Ω so that the density of protons to which it is exposed is $1/\Omega$. Indeed, we can substitute for the factor $1/\Omega$ in (152) the density n of protons. We find in this way a formula for the inverse lifetime of an anti-proton surrounded by protons with density n leading to the emission of two photons:

$$\frac{1}{\tau_{em}} = \frac{\pi e^4 n}{M^2 c^3}. \qquad (153)$$

The process of annihilation of a nucleon and an anti-nucleon by emission of a pair of pions follows essentially the same pattern and will not be calculated in detail. The result differs from (152) because in place of the electric charge e there appears the coupling constant e_2 of the Yukawa theory. There is also a difference in numerical factors arising in part from having used the Heavyside units for the Yukawa theory and not for the electromagnetic theory. The result is

$$\frac{1}{\tau_y} = \frac{e_2^4 n}{16\pi M^2 c^3}. \qquad (154)$$

Notice that the electromagnetic annihilation (153) applies only to protons and anti-protons, whereas (154) may lead to the annihilation of any nucleon-anti-nucleon pair. For example, it may lead to the annihilation of a proton and an anti-neutron with emission of either a positive and a neutral pion or of a neutral and a negative pion.

Formula (154) leads to a much higher rate of annihilation than does (153), primarily because the coupling constant e_2 is much larger than the electromagnetic coupling constant e. Indeed, it is probable that the rate of annihilation with pion emission is even faster than indicated by (154), since the total energy available in the annihilation process is sufficient to produce more than two pions so that other processes could be operative, leading to a higher over-all probability of transition.

In a numerical estimate of the rates of transition (153) and (154) the density n of the nucleons will be taken equal to that inside a nucleus which is approximately 7×10^{37}. The electromagnetic rate of decay turns out to be then $1/\tau_{em} = 1.5 \times 10^{17}$. The two-pion decay according to (154) is about one thousand times faster: $1/\tau_y = 1.5 \times 10^{20}$. For the reasons previously quoted, this value is probably an underestimate. From it would follow, for example, that a negative proton traversing a nucleus of diameter 10^{-12} cm with velocity comparable to c would have a probability of less than 1 per cent of being annihilated. Multiple pion production may raise this estimate by a large factor.

24. DECAY OF THE NEUTRAL PION INTO TWO PHOTONS

The neutral pion is supposed to be an extremely unstable particle which decays spontaneously into two photons. Experimentally its lifetime appears to be shorter than about 10^{-13} seconds. In this section the theory of the disintegration of this particle will be outlined.[4]

4. The existence of the neutral pion was suggested in attempts to understand the features of the soft component of cosmic radiation. Speculations on the two photon decay process were made in this connection by Oppenheimer, Christy, Finkelstein, and others. The first definite experimental indication of the existence of this particle came with the experiments of York and Moyer who observed the gamma rays, presumably due to the disintegration of the neutral pions, originated in nucleon collisions. Subsequently gamma rays of the same origin were observed in cosmic-ray

The process is supposed to involve a first step in which the neutral pion is changed into a virtual pair of a proton and an anti-proton. The possibility of this step is essentially contained in the Yukawa reaction (38c) since instead of a proton on the right-hand side of the equation we can write an anti-proton on the left-hand side. The virtual pair will later on disintegrate into two photons according to the electromagnetic process investigated in the previous section. We found there that for a real proton-anti-proton pair the disintegration into pions is more probable than that into photons. In the present case, however, the disintegration into pions cannot occur on account of the conservation of energy, and only the electromagnetic possibility remains. The over-all process will therefore occur in three steps, as follows:

$$\Pi^0 \to P_1 + \bar{P}_1 \to P_2 + \bar{P}_1 + \gamma_1 \to \gamma_1 + \gamma_2 . \qquad (155)$$

The last two steps are quite similar to the two steps of reaction (149), except that the energetic conditions are different since the photons emitted in this case have energy $\mu c^2 / 2$ whereas in (149) they had a much higher energy, namely, Mc^2. Process (155) is a third approximation process in which the Yukawa interaction (first step) operates once and the electromagnetic interaction (second and third steps) operates twice. The apparent matrix element for a process of this type can be calculated by a formula entirely similar to (61), namely

$$\mathcal{H}'_{no} = \sum_{km} \frac{\mathcal{H}_{nk} \mathcal{H}_{km} \mathcal{H}_{mo}}{(W_o - W_k)(W_o - W_m)} \qquad (156)$$

where zero and n are the indices of the initial and the final state, m and k are the indices of the two intermediate states, and the sum extends to all possible combinations of intermediate states. In order to evaluate this expression we need the matrix elements for the three steps. The first is the Yukawa matrix element (42) in which μc^2 will be substituted for w_s. The remaining two matrix elements due to the electromagnetic interaction are calculated from (35). In this case, however, w is the energy of the emitted photons, namely, $\mu c^2 / 2$. Both energy differences in the denominator

---

stars by Bradt and Peters and by Schein and Lord. The fact that two photons are emitted in the decay of a neutral pion was observed by Steinberger on neutral pions produced by high energy photons.

of (156) are of the order of $2 Mc^2$ since a nucleon pair has been formed.

In conclusion, the general term of the sum in (156) will be of the order of magnitude

$$\frac{\pi}{\sqrt{2}} \frac{e_2 e^2 \hbar^3}{\Omega^{3/2} \mu^{3/2} M^2 c^4}. \tag{157}$$

The sum (156) extends to an infinite number of terms because the proton P_1 may have an arbitrary momentum since momentum conservation requires only that P_1 and $\bar{P}_1$ should have equal and opposite momenta. Indeed, it is found that the sum is divergent, and a number of procedures have been suggested in order to circumvent this difficulty. This, however, cannot be done without some arbitrariness.

We expect to obtain the right order of magnitude by a crude cut-off procedure that consists in including in the sum (156) only momenta of the proton and anti-proton P_1 and $\bar{P}_1$ less than Mc. Disregarding the spin this gives a number of intermediate states

$$\frac{\Omega}{8\pi^3\hbar^3} \frac{4\pi}{3} (Mc)^3 \tag{158}$$

The order of magnitude of the effective matrix element (156) will be taken as the product of (157) and (158), namely

$$\mathcal{K}'_{no} = \frac{e_2 e^2 M}{6\sqrt{2} \pi\mu^{3/2} c\sqrt{\Omega}}. \tag{159}$$

The calculation of the lifetime for the disintegration of the neutral pions proceeds now according to the usual pattern (55) and (60). In this last formula p is the momentum of the emitted photons equal to $\mu c/2$ and v_1 and v_2 are both equal to c. The result is

$$\frac{1}{\tau} = \frac{e_2^2 e^4 M^2}{576\pi^3\hbar^4\mu c} \approx 10^{17} \text{ sec}^{-1}. \tag{160}$$

According to this theory, therefore, the lifetime of the neutral pion should be of the order of magnitude of 10^{-17} seconds, well below the experimental upper limit of about 10^{-13} seconds. The considerable uncertainty of this theory should be stressed, primarily the arbitrary cut-off introduced in evaluating the sum in (156).

25. STATISTICAL THEORY OF
PION PRODUCTION

In the previous sections the inadequacies of the
perturbation theory in handling Yukawa processes have been
stressed. They arise in part from the large value of the development
parameter and lead frequently to situations in which higher approximations give larger results than do lower approximations.
In this section a different approach to the solution of the Yukawa
problems, in which use is made of the large transition probability
between various states, will be discussed. The proposed method
may be especially useful in handling cases in which many particles
are produced (multiple processes). Examples of processes in which
several pions and perhaps also some anti-nucleons are formed are
observed in the collision of very high energy cosmic-ray nucleons.

Briefly the idea is the following. When two fast nucleons collide
their kinetic energy will be suddenly released in a small volume
surrounding the two nucleons. Since the two nucleons are surrounded by a pion field which extends up to a distance of the
order $\hbar/\mu c$, we may expect that a high concentration of energy will
be established within a volume of this size. The energy concentration lasts, of course, a very short time and is rapidly converted
into pions or other similar particles that fly away with a speed
approximating that of light. During the short time that the
energy is concentrated, however, a number of reactions involving
creation or destruction of pions may take place. In the present
discussion we want to explore the extreme assumption that a
statistical equilibrium is reached and that the probability that a
certain number of particles are created in the reaction is essentially
determined by statistical laws. This assumption would be a good
approximation if the processes of creation and destruction were
extremely fast so that they could reach statistical equilibrium
before the energy concentration disassembles. If this is not the
case one might expect a statistical theory to give an upper limit to
the number of particles formed in the collision. It seems plausible
that the approximation here proposed may give a rather good
description of collisions with very great energy since in this case

each state can be reached in a variety of ways and equilibrium should therefore be established rather quickly.

In order to specify the assumptions more completely we introduce first the volume V in which the energy transformations are supposed to take place. This volume will be the only adjustable parameter in the proposed theory. Its dimensions are of the order of magnitude $\hbar/\mu c$. One finds reasonably good agreement with the rather limited amount of experimental information by assuming that except for the Lorentz contraction V is the volume of a sphere of radius $R = \hbar/\mu c$. If W is the energy of the two colliding nucleons in the center-of-mass system the Lorentz contraction factor will be $2Mc^2/W$. We will assume therefore

$$V = V_0 \frac{2Mc^2}{W} \tag{161}$$

and

$$V_0 = \frac{4\pi}{3} R^3 = \frac{4\pi}{3} \left(\frac{\hbar}{\mu c}\right)^3. \tag{162}$$

The choice (162) is of course arbitrary. A larger volume would make phenomena of higher multiplicity more prominent. The collision of two nucleons may lead to a variety of states. Their choice is limited by several conservation theorems. Since we operate in the center-of-mass system, the total momentum of any possible final state must vanish and its total energy must, of course, be equal to the total energy W available. Other conservation theorems are the conservation of electric charge and of angular momentum. Besides the production of a number of pions, the possibility of formation of nucleon-anti-nucleon pairs also will be discussed. This process will maintain constant the difference between the final number of nucleons and anti-nucleons, which in a collision of two nucleons is, of course, two.

Statistical equilibrium means that the probability that one of the allowable states is formed in the reaction is proportional to its statistical weight. This is assumed to be proportional to the probability that all its particles are present at the same time inside the volume V. For example, the statistical weight $S(n)$ of n completely independent spinless particles with momenta p_1, p_2 $\cdots$ p_n is obtained as follows. Let $Q(W)$ be the volume of the $3n$-

dimensional momentum space inside the energy surface W. In a normalization volume Ω the number of states per unit energy interval is

$$\frac{dN}{dW} = \left(\frac{\Omega}{8\pi^3\hbar^3}\right)^n \frac{dQ(W)}{dW} . \tag{163}$$

Each particle has the probability V/Ω of being inside V. The probability that the n independent particles are all inside V is $(V/\Omega)^n$. The statistical weight is the product of this factor times (163), namely

$$S(n) = \left(\frac{V}{8\pi^3\hbar^3}\right)^n \frac{dQ(W)}{dW} . \tag{164}$$

When momentum conservation is taken into account, however, the total momentum vanishes in the center-of-mass system. The momentum space $Q(W)$ becomes therefore $3(n-1)$-dimensional and the exponent of $V/8\pi^3\hbar^3$ in (164) may also be taken to equal $n - 1$. For example, in the case of two particles one may use (60), substituting in it V for Ω:

$$S(2) = \frac{Vp^2}{2\pi^2\hbar^3(v_1 + v_2)} . \tag{165}$$

For three statistically independent nonrelativistic particles with resultant momentum zero, one finds

$$S(3) = \frac{V^2}{16\pi^3\hbar^6} \left(\frac{m_1 m_2 m_3}{m_1 + m_2 + m_3}\right)^{3/2} T^2 \tag{166}$$

where m_1, m_2, m_3 are the masses and T is the kinetic energy of the three particles.

The cross section for an event in which certain particles are produced is obtained by multiplying the total collision cross section by the probability of the event in question, normalized to one. The total cross section at high energy is assumed to be equal to the cross section of the pion cloud and will be taken to be about

$$\pi R^2 = \pi(\hbar/\mu c)^2 = 6 \times 10^{-26} \text{ cm}^2. \tag{167}$$

As a first example, the collision of two nucleons with energy barely above the threshold for the production of one pion will be considered. In this case the statistical competition is between two types of states only. One corresponds to the elastic scattering in which no pion is formed and the two nucleons escape with energy equal to that before the collision. The second possibility is that a pion is formed, in which case three particles, the two nucleons and the pion, will escape with very low kinetic energy. We shall simplify the discussion of this example by disregarding the spin and the various possible electric charges of the nucleons and the pion.

The statistical weight $S(3)$ of the three-particle state is, according to (166),

$$S(3) = \frac{V^2}{16\pi^3\hbar^6} \left(\frac{M^2\mu}{2M + \mu}\right)^{3/2} T^2 \approx \frac{V^2 M^{3/2} \mu^{3/2} T^2}{32\sqrt{2}\ \pi^3\hbar^6} \quad (168)$$

where μ has been neglected with respect to $2M$ in the denominator. In order to obtain the cross section for the three-particle process the probability of the competing process of elastic scattering must be computed. The angular momentum conservation will first be neglected. In this case the statistical weight is given by (165). Since the two nucleons are non-relativistic one finds

$$S(2) = \frac{VM^{3/2}}{4\pi^2\hbar^3} \sqrt{\mu c^2 + T}. \quad (169)$$

Here $\mu c^2 + T$ is the kinetic energy of the two nucleons in the center-of-mass system when no pion is formed.

We now partition the total cross section (167) in parts proportional to the two relative probabilities (168) and (169) in order to find the partial cross sections for elastic scattering and for pion production. Near the threshold (168) is very small compared to (169) and we can therefore assume that the probability of pion formation is given merely by the ratio of (168) to (169). This last formula can be simplified further by putting in it $T = 0$. Since the two colliding nucleons have nonrelativistic velocity we can substitute V_0 given by (162) for V. One finds in this way the total cross section for pion production,

$$\sigma_\pi = \pi \left(\frac{\hbar}{\mu c}\right)^2 \frac{1}{6\sqrt{2}} \left(\frac{T}{\mu c^2}\right)^2. \quad (170)$$

In deriving this formula the angular momentum conservation has been neglected. In this case, however, it can be taken into account easily by observing that when a pion is formed the three particles escape in a state of zero angular momentum because their kinetic energy T is very small. The competition with the scattering states will be restricted for this reason to states of zero angular momentum only. With this restriction the statistical weight $S(2)$ is, of course, smaller than (169), which includes all states. Instead of (169), one finds for s-states only

$$S(2) = \frac{R}{\pi \hbar v} = \frac{1}{\pi \mu c v} = \frac{1}{2 \pi \mu c} \sqrt{\frac{M}{\mu c^2 + T}} \qquad (171)$$

where one has assumed for the two colliding nucleons $\lambda \ll \hbar/\mu c$.

On the other hand the total cross section also must be restricted to states of zero angular momentum and is therefore smaller than (167). From the collision theory it is known that the maximum possible total cross section of the s waves is

$$\pi \lambda^2 = \pi \hbar^2/p^2 = \frac{\pi \hbar^2}{M(\mu c^2 + T)} \cdot \qquad (172)$$

The cross section for pion production is the product of (172) times the ratio of (168) and (171). Again neglecting T in (171) and (172) and substituting (162) for V, one finds

$$\sigma_\pi = \pi \left(\frac{\hbar}{\mu c}\right)^2 \frac{1}{9\sqrt{2}} \left(\frac{T}{\mu c^2}\right)^2. \qquad (173)$$

This formula differs from (170) by only a factor 2/3 which represents the angular-momentum correction.

Equation (173) should be compared with (108) which gives the cross section for the same process calculated with the perturbation theory in Section 18. This formula too should be simplified by neglecting μ with respect to M and by substituting for v its value at the threshold, $v = 2c\sqrt{\mu/M}$. One recognizes that the two formulas give the same dependence upon the energy and that (173) and (108) become identical provided the constant f has the following value:

$$f = \frac{4\sqrt{2}\,\pi^{3/2}\,\hbar^{9/2}}{3M\mu^{5/2}c^{5/2}} = 5.2 \times 10^{-62} \text{ ergs} \times \text{cm}^{9/2}.$$

The numerical value is quite close to the value 6.5×10^{-62} which has been found in Section 19. Indeed, the agreement would even be improved by taking into account the charge of the particles.

This agreement, although probably at least in part accidental, may give one some confidence that the statistical approach is allowable in calculating phenomena of this type.

26. COLLISIONS OF EXTREMELY HIGH ENERGY PARTICLES

When two nucleons collide with extremely high energy the statistical computation of the type and number of particles emitted can be further simplified by substituting for the detailed statistical analysis a thermodynamical approximation. We introduce a temperature τ to which the small volume V is heated by the sudden release of the energy W of the two nucleons. At this extremely high temperature the pions will reach a thermodynamical equilibrium similar to that of the black-body radiation. If one assumes that not only pions but also nucleons and anti-nucleon pairs may be formed, one may assume also that these particles will reach a similar equilibrium. All other processes, for example the emission of photons, are far too slow to develop.

Since the energy and the temperature are extremely high we shall further assume that all particles are extreme relativistic in the center-of-mass system. For each particle, therefore, the relationship between energy and momentum will be, as for a photon $w = cp$. Since the pions, like the photons, obey the Bose-Einstein statistics, the pion gas will build up to an energy density given by a formula almost identical to Stefan's law of radiation. The only difference is in the statistical weight which is assumed here to be three for the pions corresponding to the three possible values of the electric charge and two for the photons corresponding to the two polarization possibilities. The energy density of the pions is obtained, therefore, by multiplying the ordinary energy density of Stefan's law by 3/2. This density is therefore

$$\frac{3 \times 6.494\tau^4}{2\pi^2\hbar^3 c^3}. \tag{175}$$

The factor $6.494 = \pi^4/15$ is six times the sum of the inverse fourth powers of the integral numbers.

Nucleons and anti-nucleons also will form a black-body-like

radiation whose energy density in the extreme relativistic case is proportional to τ^4. The coefficient is different for two reasons. The statistical weight should be taken equal to 8 because each particle has two possible spin orientations, and there are four types of nucleons. Also a small difference is due to the fact that nucleons obey the Pauli principle instead of the Bose-Einstein statistics. Taking these factors into account one finds that the energy density of the nucleons is

$$\frac{4 \times 5.682\tau^4}{\pi^2 \hbar^3 c^3}. \tag{176}$$

The numerical factor is $5.682 = 6 \Sigma_1^\infty (-1)^{n+1}/n^4$. The temperature τ is computed by equating the total energy W available to the product of the sum of the two energy densities (175) and (176) by the volume V. One finds, using (161):

$$\tau^4 = .152 \frac{\hbar^3 c^3 W^2}{Mc^2 V_0}. \tag{177}$$

In the extreme relativistic case the densities n_π and n_N of the pions and the nucleons are proportional to the third power of the temperature and are given by

$$n_\pi = .367 \ \tau^3/\hbar^3 c^3; \qquad n_N = .855 \ \tau^3/\hbar^3 c^3. \tag{178}$$

The numbers of the two kinds of particles are obtained by multiplying these densities by the volume V.

The result, however, must be corrected because the angular-momentum conservation has been disregarded. The angular-momentum conservation modifies the statistics in the sense of increasing the occupation of states whose angular momentum is parallel to that of the two original colliding nucleons. This has two effects. One is to change the formula for the number of particles by a factor which has been computed numerically to be .51. The second is to make the angular distribution of the particles in the center-of-gravity system non-isotropical and such as to contain a sizable concentration of particles moving in directions forming small angles with the directions of the two original nucleons. Including the angular-momentum correction, one finds the following formula for the numbers of pions and nucleons emerging out of an extremely high-energy collision:

$$\text{Number of pions} = .091 \left(\frac{V_0 M W^2}{c\hbar^3}\right)^{1/4}$$

$$= .54\sqrt{W/Mc^2} = .64\,(W'/Mc^2)^{1/4} \qquad (179)$$

$$\text{Number of nucleons} = .21 \left(\frac{V_0 M W^2}{c\hbar^3}\right)^{1/4}$$

$$= 1.3\sqrt{W/Mc^2} = 1.5\,(W'/Mc^2)^{1/4} \qquad (180)$$

where W' is the energy of the bombarding nucleon in the laboratory system.

In experiments with photographic plates one observes only the charged particles. From the previous data it follows that this number should be equal to

$$1.2\,(W'/Mc^2)^{1/4}. \qquad (181)$$

This theory leads to the expectation that a considerable number of anti-protons should be formed at high energy. This number should be given by $1/4$ of the number (180) and should be therefore

$$.38\,(W'/Mc^2)^{1/4}. \qquad (182)$$

The validity of this formula, however, is restricted to energies well above 10^{13} ev for the bombarding particle. At lower energies the temperature is not high enough to yield a large probability of nucleon-anti-nucleon pair formation. If one discusses with the statistical method here proposed the probability of anti-nucleon formation at lower energies one finds that for bombarding energy of the order of 10 Bev the number of anti-nucleons formed is of the order of a few thousandths of the number of pions formed.

In the intermediate energy range between bombarding energies 10^{10} and 10^{12} ev, one is allowed therefore to disregard in first approximation the pair formation and only the pion black-body radiation should be considered. For the total number of pions one finds then instead of (179) the following approximate formula

$$\text{Number of pions} = 1.34\,(W/Mc^2 - 2)^{3/4}/(W/Mc^2)^{1/4}. \qquad (183)$$

One-third of these pions should be neutral.

Recently strong evidence has been brought forth to indicate the probable existence of a heavy meson. If these heavy mesons are bound to nucleons by an interaction comparable in strength to that of the pions they also should be formed in high-energy nuclear collisions and they would compete statistically with the pions. On account of their greater mass the competition, however would be favorable to the pions at low energies. At very high energies the competition would instead be determined by the spin of the heavy mesons.

APPENDICES

Quantization of the Radiation Field

In this appendix a somewhat more detailed account of the quantization of the radiation field is given. The discussion is restricted to the radiation field and does not include the general electromagnetic field. By radiation field is meant the electromagnetic field obtained by superposition of transversal electromagnetic waves. In addition a general electromagnetic field would include, for example, the Coulomb forces. The radiation field contained in a cavity of volume Ω can be analyzed as a superposition of plane waves. It is convenient at first to make the analysis in terms of standing waves rather than progressive waves. We distinguish the various standing waves by an index s.

A general electromagnetic field can be described with a scalar and a vector potential. For a pure radiation field, however, one may always assume that only the vector potential A is different from zero. The standing wave number s can then be represented by its vector potential

$$A_s = \epsilon_s q_s \cos f_s \cdot r. \tag{1}$$

Here f_s is the vector wave number multiplied by 2π; ϵ_s is a unit vector perpendicular to f_s and pointing in the polarization direction; q_s is the instantaneous amplitude of the standing wave. In order to keep the notation simple the phase of this particular wave has been taken as zero. With this analysis the radiation field is completely described by the quantities q_s that may be regarded as the coordinates of the physical system represented by the radiation field. They are functions of the time which, as long as the radiation field is not disturbed, will vary sinusoidally with time with a frequency corresponding to the wave number f_s.

The electric and magnetic fields E_s and H_s of the wave (1) are immediately computed with the relationships

$$E = - \tfrac{1}{c}\dot{A}; \qquad H = \nabla \times A. \tag{2}$$

They are

$$E_s = - \tfrac{\epsilon_s}{c}\dot{q}_s \cos f_s \cdot r; \qquad H_s = (\epsilon_s \times f_s)q_s \sin f_s \cdot r. \tag{3}$$

The total energy of the wave (1) is given by

$$\mathcal{H}_s = \frac{\Omega}{8\pi}\,(\overline{E_s^2} + \overline{H_s^2}) = \frac{\Omega}{16\pi c^2}\,\dot{q}_s^2 + \frac{\Omega f_s^2}{16\pi}\,q_s^2. \tag{4}$$

By making use of the orthogonality of the functions $\cos f_s \cdot r$ one can prove readily that the total energy of the radiation field is given by $\Sigma \mathcal{H}_s$ where the sum is extended to all modes s. In other words, each mode contributes independently to the total energy.

Formula (4) may be identified with the well-known expression

$$\frac{m}{2}\,\dot{q}_s^2 + \frac{m}{2}\,\omega_s^2 q_s^2 \tag{5}$$

for the energy of an oscillator of mass m and frequency ω_s. The electromagnetic wave s behaves like an oscillator whose coordinate is q_s and whose mass and frequency are

$$m = \frac{\Omega}{8\pi c^2}; \qquad \omega_s = c \,|\, f_s \,|. \tag{6}$$

The frequency ω_s is what one would expect for a wave of wave number f_s traveling with the velocity of light. One might be puzzled by the fact that the dimensions of m are not those of a mass. This is due to the fact that the variable q_s does not have the dimensions of a length.

The radiation field can now be quantized as an assembly of the oscillators s. Their energy is given as in the formula (2, text) by

$$W = \Sigma\, \hbar\omega_s n_s \tag{7}$$

where the constant zero-point energy has been omitted. As explained in the text the numbers n_s are interpreted as the number of photons present in each mode.

The variable q_s is an operator that has matrix elements connecting states for which the number of photons n_s changes by ± 1. These matrix elements are obtained by substituting in the oscillator formula (4, text) the value (6) of m. They are

$$q_s(n_s \to n_s - 1) = \sqrt{\frac{4\pi\hbar c^2}{\Omega\omega_s}} \sqrt{n_s}$$

$$q_s(n_s \to n_s + 1) = \sqrt{\frac{4\pi\hbar c^2}{\Omega\omega_s}} \sqrt{n_s + 1}$$

(8)

The vector potential $A(r)$ at the position r is the sum of expressions like (1) which are proportional to q_s. In this sum only the quantities q_s are operators. Consequently the vector potential $A(r)$ also will be an operator. Note that the vector r which specifies a position in space is not an operator. For each position r there is an operator that represents the vector potential $A(r)$ observable at that position. The operator $A(r)$ has matrix elements connecting the same states as do the operators q_s. $A(r)$ therefore causes changes of the number n_s of photons by ± 1. The matrix elements are obtained by multiplying the matrix elements (8) by the coefficient of q_s in the expression of A; namely by $\epsilon_s \cos f_s \cdot r$. In this way one finds the matrix elements of the operator $A(r)$:

$$A(r)(n_s \to n_s - 1) = \epsilon_s \cos f_s \cdot r \sqrt{\frac{4\pi\hbar c^2}{\Omega\omega_s}} \sqrt{n_s}$$

$$A(r)(n_s \to n_s + 1) = \epsilon_s \cos f_s \cdot r \sqrt{\frac{4\pi\hbar c^2}{\Omega\omega_s}} \sqrt{n_s + 1}$$

(9)

Note that both $A(r)$ and its matrix elements are vectors.

The formulas (9) are similar but not identical to the formulas (5, text). The difference arises from the fact that the radiation field has been analyzed here in standing rather than progressive waves. Standing waves can be considered as a superposition of two progressive waves traveling in opposite directions. Consequently the photon analysis mixes photons traveling in the direction of the vector f_s and in the opposite direction. The standing-wave analysis presented here has, however, the advantage of being more elementary because it avoids the use of non-Hermitian operators. The analysis in terms of progressive waves follows.

Classically a progressive wave may be described as a superposition of two standing waves with a phase shift of a quarter-wave length. This leads one to expect also that in the quantum mechanical radiation theory a progressive wave will be obtained by combining in a suitable form two standing waves with a quarter-wave-phase shift. It will be found that the photons of these two degrees of freedom of the radiation field can be analyzed into one set of photons moving in the positive and one in the negative direction. We simplify the notation by assigning the index 1 to the oscillator (1) and the index 2 to the oscillator representing a standing wave shifted by a quarter wave from (1). For these two oscillators the vector potentials will be

$$A_1 = \epsilon q_1 \cos f \cdot r \qquad A_2 = \epsilon q_2 \sin f \cdot r. \tag{10}$$

No index has been given to ϵ and f since they are the same for both oscillators. Also the mass and frequency of these oscillators are the same and are given by (6). Besides the coordinates q_1 and q_2 of the two oscillators the conjugate momenta $p_1 = m\dot{q}_1$ and $p_2 = m\dot{q}_2$ will also be introduced. Each oscillator contributes to the total energy a term like (5). When the momenta are introduced the contribution of the two oscillators to the total energy is

$$\mathcal{H} = \frac{1}{2m}(p_1^2 + p_2^2) + \frac{m\omega^2}{2}(q_1^2 + q_2^2) \tag{11}$$

which has the energy levels

$$\mathcal{H} = \hbar\omega(n_1 + n_2) \tag{12}$$

where again the zero-point energy has been omitted.

The operator q_1 induces transitions from n_1 to $n_1 \pm 1$. The same transitions are also induced by the operator p_1. The matrix elements of q_1 and p_1 are:

$$q_1(n_1 \to n_1 \pm 1) = \sqrt{\frac{\hbar}{2m\omega}} \begin{cases} \sqrt{n_1 + 1} \\ \sqrt{n_1} \end{cases}$$

$$p_1(n_1 \to n_1 \pm 1) = \pm i \sqrt{\frac{\hbar m\omega}{2}} \begin{cases} \sqrt{n_1 + 1} \\ \sqrt{n_1} \end{cases}. \tag{13}$$

It is convenient to introduce in place of q_1 and p_1 two linear combinations a_1 and a_1^* as follows:

$$a_1 = \sqrt{\frac{m\omega}{2\hbar}}\, q_1 + \frac{i}{\sqrt{2\hbar m\omega}}\, p_1$$
$$a_1^* = \sqrt{\frac{m\omega}{2\hbar}}\, q_1 - \frac{i}{\sqrt{2\hbar m\omega}}\, p_1 . \tag{14}$$

These two operators are not Hermitian. Their matrix elements can be obtained by combining the matrix elements (13), and one verifies immediately that the operator a_1 has matrix elements different from zero only for transitions from n_1 to $n_1 - 1$, while a_1^* has matrix elements different from zero only for transitions from n_1 to $n_1 + 1$. The matrix elements computed from (13) and (14) are

$$a_1(n_1 \to n_1 - 1) = \sqrt{n_1}$$
$$a_1^*(n_1 \to n_1 + 1) = \sqrt{n_1 + 1}. \tag{15}$$

The operators a and a^* are called destruction and creation operators since a induces destructive transitions in which one photon disappears and a^* induces creative transitions in which one photon appears. From the matrix representation (15) one can also derive the exchange relationship of these two operators

$$a_1 a_1^* - a_1^* a_1 = 1 \tag{16}$$

which is equivalent to the exchange relationship $pq - qp = \hbar/i$ of the operators p and q. From (15) one finds also a formula

$$a_1^* a_1 = n_1 . \tag{17}$$

The quantity $a_1^* a_1$ has, therefore, integral eigenvalues that represent the number of photons of the oscillator 1. Formula (14) can be inverted and q_1 expressed in terms of a_1 and a_1^* as follows:

$$q_1 = \sqrt{\frac{\hbar}{2m\omega}}\, (a_1 + a_1^*). \tag{18}$$

In a similar manner one introduces a creation and a destruction operator a_2 and a_2^* for the oscillator 2. The vector potential A

contributed by the two degrees of freedom is the sum of A_1 and A_2. Using (10) and (18) one finds

$$A = A_1 + A_2 = \epsilon \sqrt{\frac{\hbar}{2m\omega}}$$

$$\cdot \{(a_1 + a_1^*) \cos f \cdot r + (a_2 + a_2^*) \sin f \cdot r\} \qquad (19)$$

$$= \frac{\epsilon}{2} \sqrt{\frac{\hbar}{m\omega}} \{(a_+ + a_-^*)e^{if \cdot r} + (a_+^* + a_-)e^{-if \cdot r}\}$$

where

$$a_+ = \frac{a_1 - ia_2}{\sqrt{2}} \qquad a_+^* = \frac{a_1^* + ia_2^*}{\sqrt{2}}$$

$$a_- = \frac{a_1 + ia_2}{\sqrt{2}} \qquad a_-^* = \frac{a_1^* - ia_2^*}{\sqrt{2}}. \qquad (20)$$

By using (17) the energy (12) of the two oscillators can be expressed as follows:

$$\mathcal{H} = \hbar\omega(a_1^* a_1 + a_2^* a_2). \qquad (21)$$

This formula can be transformed by introducing a_+ and a_- in place of a_1 and a_2. It becomes

$$\mathcal{H} = \hbar\omega(a_+^* a_+ + a_-^* a_-) = \hbar\omega(n_+ + n_-) \qquad (22)$$

where

$$n_+ = a_+^* a_+ \qquad n_- = a_-^* a_-. \qquad (23)$$

Formula (22) can be immediately obtained from the definition of a_+ and a_-, taking into account that a_1 and a_1^* commute with a_2 and a_2^* since they are functions of the coordinates and momenta of two different oscillators. One also verifies readily from the definition that a_+ and a_- have the same exchange relationships as a_1 and a_2, for example:

$$a_+ a_+^* - a_+^* a_+ = 1$$

$$a_+ a_- - a_- a_+ = 0. \qquad (24)$$

The operators a_1 and a_2, or any linear combinations of them like a_+ or a_-, induce a transition in which the total number of photons in the two degrees of freedom decreases by one unit. Similarly, a_1^* and a_2^* or any linear combinations of them like a_+^* or a_-^* are operators that induce an increase by one in the number of photons of the two degrees of freedom. By combining the matrix elements (15) and the similar matrix elements of a_2 and a_2^*, one finds that the operators a_+ and a_+^* operate on the number n_+ as a_1 and a_1^* operate on the quantum number n_1.

The numbers n_+ and n_- may be interpreted as the number of photons traveling in the positive and negative directions. This is best seen by computing the contribution of our two degrees of freedom to the total electromagnetic momentum:

$$\mathfrak{M} = \int \frac{E \times H}{4\pi c} \, d\Omega. \tag{25}$$

Using (2) one finds with a fairly easy calculation

$$\mathfrak{M} = \eta \frac{\hbar\omega}{c} (n_+ - n_-) \tag{26}$$

where η is a unit vector in the propagation direction. This formula together with (22) indicates that to each photon n_+ one associates the energy $\hbar\omega$ and the momentum $\hbar\omega/c$ in the positive direction. To each photon n_- one associates the same energy and opposite momentum.

The vector potential A given by (19) has the same operatorial properties as the creation and destruction operators in terms of which it is expressed. In particular the operator $A(r)$ will have matrix elements inducing creation and destruction of the photons n_+. They are given by the matrix elements of a_+ and a_+^* multiplied by the coefficient of these quantities in the expression of A, namely

$$\begin{aligned}
A(r)(n_+ \to n_+ - 1) &= \frac{\epsilon}{2} \sqrt{\frac{\hbar}{m\omega}} \, e^{if \cdot r} \sqrt{n_+} \\
A(r)(n_+ \to n_+ + 1) &= \frac{\epsilon}{2} \sqrt{\frac{\hbar}{m\omega}} \, e^{-if \cdot r} \sqrt{n^+ + 1}.
\end{aligned} \tag{27}$$

These expressions are essentially identical to (5, text) to which they reduce by introducing in them the value (6) of the mass.

Second Quantization
with Pauli Principle

In this appendix the field theory of electrons or that of any particle obeying the Pauli principle will be discussed. For simplicity spin and relativity will be neglected. The field ψ whose photons are the electrons will then obey the Schroedinger equation (20, text):

$$\frac{\partial \psi}{\partial t} = \frac{i\hbar}{2m} \nabla^2 \psi. \tag{1}$$

Together with this equation its complex conjugate

$$\frac{\partial \psi^*}{\partial t} = -\frac{i\hbar}{2m} \nabla^2 \psi^* \tag{2}$$

should also be considered. The energy density of the field ψ must have the property that its volume integral over the space Ω is a time constant as a consequence of the field equations (1) and (2).

One finds that a suitable expression that has this property is $(\hbar^2/2m)\nabla\psi^* \cdot \Delta\psi$. The total energy is then given by

$$H = \frac{\hbar^2}{2m} \int \nabla\psi^* \cdot \nabla\psi \, d\Omega = -\frac{\hbar^2}{2m} \int \psi^* \nabla^2 \psi \, d\Omega. \tag{3}$$

Note that the two integral expressions can be transformed into each other by partial integrations. It is assumed that ψ and ψ^* behave at great distances in such a way that the surface integrals arising from the partial integrations vanish.

It will now be assumed that $\psi(r)$ and $\psi^*(r)$ at a given position in space, r, are operators. These quantities therefore will have in general a non-commutative multiplication law.

From the general principles of quantum mechanics it follows

that any physical quantity A represented by an operator which does not explicitly contain the time varies with time according to the differential equation

$$\frac{dA}{dt} = \frac{i}{\hbar}(HA - AH) \tag{4}$$

where H is the total energy operator of the system. This property will apply to $\psi(r)$ and $\psi^*(r)$ and can be used in order to calculate the time derivatives of these quantities. These derivatives, on the other hand, are given by the field equations, (1) and (2). From the comparison one obtains

$$\nabla^2 \psi = \frac{2m}{\hbar^2}(H\psi - \psi H)$$

$$\nabla^2 \psi^* = -\frac{2m}{\hbar^2}(H\psi^* - \psi^* H). \tag{5}$$

The choice of exchange relations of ψ and ψ^* is limited by these equations in which for H one should insert the expression (3).

One can verify readily that two simple types of exchange relations of the field quantities at two positions r and r' in space are consistent with this requirement. One of them is the following:

$$\psi(r)\psi(r') + \psi(r')\psi(r) = 0$$

$$\psi^*(r)\psi^*(r') + \psi^*(r')\psi^*(r) = 0 \tag{6}$$

$$\psi(r)\psi^*(r') + \psi^*(r')\psi(r) = \delta(\vec{r}' - \vec{r}).$$

The other possibility differs from (6) because in it the plus signs are replaced by minus signs. This second possibility would lead to a quantization of the type discussed in Appendix 1 yielding the Bose-Einstein statistics and will not be considered here. The possibility (6) leads instead to the exclusion principle. We assume therefore that our field operators ψ and ψ^* have the commutation rules (6). They express an anti-commutative property of the field quantities because from them it follows that the operators ψ and/or ψ^* anti-commute when taken at two different points, r and r', in space.

In order to avoid the formal difficulties arising from the fact that there is a continuous infinity of operators $\psi(r)$ and $\psi^*(r)$ corresponding to the continuous infinity of the points in space, it is convenient to develop ψ and ψ^* according to a complete system of eigenfunctions $u_s(r)$. In most cases the eigenfunction u_s will be normalized plane waves and one will then have

$$u_s(r) = \frac{1}{\sqrt{\Omega}}\, e^{(i/\hbar)p_s \cdot r} \tag{7}$$

in which the allowable values of the momentum p_s are given by (17, text). In other cases it is expedient to choose the u_s's differently. For example, when a system of time-independent external forces acts on the electron it is usually convenient to take the u_s's to be the eigenfunctions of the Schroedinger problem in the given field of external forces.

In any case we will now develop $\psi(r)$ and $\psi^*(r)$ as follows:

$$\psi(r) = \sum_s a_s u_s(r); \qquad \psi^*(r) = \sum_s a^* u_s(r). \tag{8}$$

Notice that in these formulas $u_s(r)$ is an ordinary number (not an operator). Instead the coefficients a_s and a_s^* are non-commutative operators. Their commutative properties can be derived by substituting (8) into (6) and making use of the orthogonality and completeness properties of the functions u_s.
They are

$$\left.\begin{aligned} a_s a_k + a_k a_s &= 0 \\ a_s^* a_k^* + a_k^* a_s^* &= 0 \\ a_s a_k^* + a_k^* a_s &= \delta_{sk}. \end{aligned}\right\} \tag{9}$$

These exchange properties of the a's are completely equivalent to the exchange properties (6) of the ψ's. The task now remains to find an actual expression of the operators a and a^* consistent with the commutation rules (9). Notice first of all that putting $k = s$ in the first equation (9) one obtains $a_s^2 = 0$ and similarly from the second equation $a_s^{*2} = 0$. It would be wrong to conclude from these two relationships that a_s and a_s^* vanish because a_s and a_s^* are non-Hermitian operators, and for this reason the fact that their squares vanish does not prove that the operators themselves are zero. The quantity $n_s = a_s^* a_s$ which is the product of two Hermitian conjugate quantities will be real. One can prove readily that its eigenvalues are 0 and 1. This is seen by computing the

square of this quantity and making use of the exchange relation
(9) as follows:

$$n_s^2 = (a_s^* a_s)^2 = a_s^* a_s a_s^* a_s = a_s^*(1 - a_s^* a_s)a_s$$
$$= a_s^* a_s - a_s^{*2} a_s^2 = a_s^* a_s = n_s.$$

Since $n_s^2 = n_s$ and n_s is Hermitian the only possible values of n_s
are the roots of this equation, 0 and 1. From the exchange relations
(9) it follows also that n_s and n_k commute. These two properties
permit one to define as a set of quantum numbers the values
(0 or 1) of all the quantities n_s. One finds that n_s represents the
number of electrons in the state defined by the eigenfunction u_s.
The fact that n_s can take only the values 0 and 1 obviously corresponds to the Pauli principle. We will make plausible the state-
ment that n_s actually represents the number of electrons in the
state s (n_s = occupation number of state s) as follows. Take for the
eigenfunctions u_s the special expression (7) and substitute (8)
in the expression of the energy (3). Making use of the fact that
$\nabla^2 u_s = -(p_s^2/\hbar^2)u_s$ and of the orthogonality and normalization of
the u_s's one finds

$$H = \sum_s \frac{p_s^2}{2m} a_s^* a_s = \sum w_s a_s^* a_s = \sum w_s n_s \qquad (10)$$

where

$$w_s = \frac{p_s^2}{2m} \qquad (11)$$

is the kinetic energy of an electron of momentum p_s. Formula
(10) is the expression of the total energy. It obviously corresponds
to a state in which there are n_s electrons in a state of energy w_s.
The same result would be found also if the electrons were acted
upon by external forces.

The operators a_s and a_s^* must now be given explicitly for the
representation of the state in terms of occupation numbers. We
will first consider only one value of s. In this case there are only
two possible states $n_s = 1$ and $n_s = 0$. Hence the matrices representing the operators a_s and a_s^* will have only two rows and two
columns.

Consider the following three Hermitian quantities:

$$\xi = a_s^* + a_s \qquad \eta = i(a_s - a_s^*) \qquad \zeta = 2a_s^* a_s - 1$$

From the commutation rules (9) it follows that these three operators have the same general properties as the Pauli spin operators, namely $\xi^2 = \eta^2 = \zeta^2 = 1$, and further that any two of the operators ξ, η, ζ anti-commute. The three quantities therefore can be represented by the well-known Pauli matrices

$$\xi = \begin{vmatrix} 0 & 1 \\ 1 & 0 \end{vmatrix} \qquad \eta = \begin{vmatrix} 0 & -i \\ i & 0 \end{vmatrix} \qquad \zeta = \begin{vmatrix} 1 & 0 \\ 0 & -1 \end{vmatrix}.$$

In this representation ζ is diagonal. Its eigenvalues ± 1 correspond to the two eigenvalues 1 and 0 of $n_s = a_s^* a_s = (\zeta + 1)/2$.

From the above matrix representations of ξ and η one finds the matrices a_s and a_s^* as follows:

$$a_s = \tfrac{1}{2}(\xi - i\eta) = \begin{vmatrix} 0 & 0 \\ 1 & 0 \end{vmatrix}; \qquad a_s^* = \tfrac{1}{2}(\xi + i\eta) = \begin{vmatrix} 0 & 1 \\ 0 & 0 \end{vmatrix}.$$

Notice that a_s and a_s^* are non-Hermitian matrices. The matrix a_s has only one element different from zero. It is the element corresponding to the transition $n_s = 1 \rightarrow n_s = 0$ (destructive transition). This element has value 1. Similarly the matrix a_s^* has only one element equal to 1 and all other elements equal to 0. The non-vanishing element corresponds to the creative transition $n_s = 0 \rightarrow n_s = 1$.

Similar results are found also when s can take many values. One can prove that the operator a_s has matrix elements different from 0 only for the transition in which n_s decreases by one unit from $n_s = 1$ to $n_s = 0$ and all the remaining occupation numbers are unchanged. It is further found that when this matrix element is non-vanishing its value is $(-1)^l$ where l is the number of occupied states with index less than s. The operator a_s is called a destruction operator because it causes only transitions in which one electron disappears. The operator a_s^* has the opposite property and is called a creation operator. Its only non-vanishing matrix elements connect states in which all the occupation numbers except n_s are unchanged and n_s increases from 0 to 1. The value of this matrix element is, as before, $(-1)^l$. One can verify with a straightforward calculation that the matrices a_s and a_s^* so defined actually obey all the exchange relations (9).

According to (8), ψ and ψ^* are linear combinations of a_s and a_s^* with non-matrix coefficients. Therefore one can write down immediately the matrix elements of the operator $\psi(r)$. It has matrix elements connecting states in which one only of the occupation numbers, for example n_s, changes from 1 to 0 while all the others are unchanged. The matrix element in question follows immediately from the previous definition of the matrix elements of a_s and is

$$(-1)^l u_s(r). \tag{12}$$

Similarly the operator $\psi^*(r)$ will have matrix elements for transitions in which one of the occupation numbers, for example n_s, changes from 0 to 1 while all the rest remain constant. The corresponding matrix element is

$$(-1)^l u_s^*(r). \tag{13}$$

Formulas (12) and (13) correspond to the matrix element (22) given in the text. With the choice (7) of the eigenfunctions u_s they give the matrix elements (21, text).

APPENDIX THREE

Measurability of the Fields

In this appendix the meaning of the components of a field will be briefly discussed from the point of view of the theory of measurement of quantum mechanics. If A and B are two real physical quantities represented by Hermitian operators, it is usually assumed that their simultaneous measurement is unrestrictedly possible only if and when the operators A and B commute.

For example, in the case of the electromagnetic field one would expect that a single component of the electric field at one point of space should be measurable. The simultaneous measurement of all the components of the electric and magnetic field, however, would be impossible because some of these components do not commute. Actually there may be difficulties even with the measurement of a single component at one point in space because the expectation value of its magnitude is infinitely large as a consequence of the zero-point energy of the field oscillators. It is found that the zero-point energy of the high-frequency components contributes an infinite field strength even in the absence of any photons. This difficulty might be partially avoided by measuring a field component with a device that loses sensitivity at very high frequencies. For example, instead of measuring the electric field at one point in space its average over a finite volume with dimensions that are small when compared with the wave lengths in which one is interested might be measured. With regard to the complementarity of the measurement of various components of the electromagnetic field, in many cases the limitation due to them is negligible. This applies for example to the field irradiated by a radio station. In this case the number of photons is so large that the occupation numbers n_s are large. One is therefore close to the classical limit in which the effect of the uncertainty relations is small.

The scalar field φ studied in Section 3 has similar properties. Except for the contribution of the zero point energy which can be handled as in the case of the electromagnetic field, it is to be expected that the field φ at a given position is measurable. A simultaneous measurement of φ and $\dot{\varphi}$ at the same place is prevented, however, by the fact that these two quantities do not commute. This limitation becomes unimportant when the number of particles in the field is so great that the classical limit is approached.

From the operatorial definition of the components of fields whose photons obey the Bose-Einstein statistics it follows that any two components at two different points in space, r and r', are represented by commuting operators. There is for this reason no further difficulty in the simultaneous measurement of components of the field at two different positions.

The situation is different for fields of particles obeying the Pauli principle. It has been seen in Appendix 2 that in this case field components at two different positions in space anti-commute instead of commuting. Therefore the measurement of a field component at one place is incompatible with a simultaneous measurement of the same or another field component at a different place. If the field amplitudes were meaningful physical quantities this would be in contradiction with relativity since the signal that carries the disturbance from the measurement at one place to that at the other should travel at infinite speed.

All this makes one wonder whether the Pauli fields are more than a mathematical symbolism. The saving feature is that in all meaningful expressions, as for example in the energy and charge densities, the amplitudes of Pauli fields always appear in monomial terms containing as factors an even number of such amplitudes. One verifies readily[1] that the values of such products at two different points always commute as a consequence of the anti-commutation of the field components. This removes the conflict with the relativity requirements.

1. Let $\alpha(r)$, $\beta(r)$ be two field components that anti-commute as follows:

$$\alpha(r_1)\alpha(r_2) + \alpha(r_2)\alpha(r_1) = 0, \qquad \beta(r_1)\beta(r_2) + \beta(r_2)\beta(r_1) = 0$$
$$\alpha(r_1)\beta(r_2) + \beta(r_2)\alpha(r_1) = 0$$

where r_1 and r_2 are two different points in space. Then

$$\alpha(r_1)(\beta r_1)\alpha(r_2)\beta(r_2) = -\alpha(r_1)\alpha(r_2)\beta(r_1)\beta(r_2) =$$
$$= \alpha(r_2)\alpha(r_1)\beta(r_1)\beta(r_2) = -\alpha(r_2)\alpha(r_1)\beta(r_2)\beta(r_1) =$$
$$= \alpha(r_2)\beta(r_2)\alpha(r_1)\beta(r_1)$$

which proves that $\alpha(r_1)\beta(r_1)$ commutes with $\alpha(r_2)\beta(r_2)$.

Relativistic Invariance

Some points of view concerning the relativistic invariance of field theories will be briefly outlined in this appendix. A point where misunderstanding may arise should first be cleared up. Relativistically, momentum and energy are space-and-time components of a four-vector. One might expect therefore that an energy counterpart of the momentum-conservation theorem of Section 7 should exist. This, however, is wrong. There is no restriction which requires that non-vanishing matrix elements should connect only states of equal energy. This lack of symmetry in the behavior of the matrix elements of the interaction energy with respect to energy and momentum is not in contradiction with the relativistic invariance of the over-all process. It arises from having used the Hamiltonian scheme. The Hamiltonian, by putting the emphasis on the energy, and the Schroedinger equation by putting the emphasis on events that occur at the same time at various points in space, are schemes that formally are not relativistically invariant. Meaningful interactions, however, are always such that the final results obey the relativity requirements. A generalization of the Schroedinger equation which is formally relativistically invariant is the Tomonaga equation. Its use has decided advantages in the study of the relativistic behavior of fields and their interactions. Since this, however, was not our primary concern in these lectures we have adhered to the more usual Schroedinger approach.

Sometimes it is advantageous to describe the properties of the fields by using, at least at first, the Lagrangian instead of the Hamiltonian. Standard rules of general mechanics permit one to proceed formally in a second step from the Lagrangian to the Hamiltonian. The advantage is that a Lagrangian density $\mathcal{L}$ which is formed by a relativistically invariant combination of components of the various fields and their space or time derivatives automatically yields relativistically invariant field equations. The field equations are the variational equations corresponding to the extremal problem

$$\delta \int \mathfrak{L} \, d\Omega \, dt = 0. \tag{1}$$

Since $d\Omega \, dt$, the volume element of space-time, is relativistically invariant, it follows also that the field equations are invariant.

In inventing a new type of interaction adapted to produce a desired reaction between elementary particles one will as a rule add to the Lagrangian a term containing products of the field amplitudes and their complex conjugates so as to obtain the proper combination of creation and destruction operators. Sometimes instead of the amplitudes their derivatives may be used. The requirement of relativistic invariance is helpful because it somewhat limits the choice of the interaction Lagrangian.

Useful properties to be remembered in this connection are the relativistic transformations of the four components of a Dirac wave function. Particularly useful is the fact that certain special bilinear combinations of the Dirac components and of their complex conjugates transform like scalars, four vectors and anti-symmetrical tensors. Here are these combinations:

$$S = \tilde{\psi}\beta\Psi \qquad \text{(scalar)} \tag{2}$$

$$P_s = \tilde{\psi}\beta\alpha_1\alpha_2\alpha_3\Psi \qquad \text{(pseudoscalar)} \tag{3}$$

$$\left\{ \begin{array}{ll} V_1 = \tilde{\psi}\alpha_1\Psi; & V_2 = \tilde{\psi}\alpha_2\Psi \\ V_3 = \tilde{\psi}\alpha_3\Psi; & V_0 = \tilde{\psi}\Psi \end{array} \right. \qquad \text{(four vector)} \tag{4}$$

$$\left\{ \begin{array}{ll} Pv_1 = \tilde{\psi}\, \dfrac{\alpha_2\,\alpha_3}{i}\, \Psi; & Pv_2 = \tilde{\psi}\, \dfrac{\alpha_3\,\alpha_1}{i}\, \Psi \\[2ex] Pv_3 = \tilde{\psi}\, \dfrac{\alpha_1\,\alpha_2}{i}\, \Psi; & Pv_0 = \tilde{\psi}\, \dfrac{\alpha_1\,\alpha_2\,\alpha_3}{i}\, \Psi \end{array} \right. \qquad \begin{array}{l} \text{(pseudo-four} \\ \text{vector)} \end{array} \tag{5}$$

$$\left\{ \begin{array}{ll} T_{23} = \tilde{\psi}\, \dfrac{\beta\alpha_2\,\alpha_3}{i}\, \Psi; & T_{31} = \tilde{\psi}\, \dfrac{\beta\alpha_3\,\alpha_1}{i}\, \Psi \\[2ex] T_{12} = \tilde{\psi}\, \dfrac{\beta\alpha_1\,\alpha_2}{i}\, \Psi; & T_{10} = \tilde{\psi}\, \dfrac{\beta\alpha_1}{i}\, \Psi \\[2ex] T_{20} = \tilde{\psi}\, \dfrac{\beta\alpha_2}{i}\, \Psi; & T_{30} = \tilde{\psi}\, \dfrac{\beta\alpha_3}{i}\, \Psi \end{array} \right. \qquad \begin{array}{l} \text{(anti-symmetric} \\ \text{tensor)} \end{array} \tag{6}$$

It should be noticed that the transformation properties of the above expressions are independent of whether ψ and Ψ are the same or different fields. For example Ψ could be the neutron field N and ψ the proton field P.

Relationships between Interaction Constants

The six interaction processes with coupling constants e, e_2, e_3, g_1, g_2, g_3 have been treated in the text as primary and independent processes. Some of them, however, may be a consequence of others.

For example, the interaction (52, text) could be a consequence of the Yukawa interaction (38, text) and of the interaction (43, text), as follows:

$$P + \mu^- \to P + \Pi^- + \bar{\nu} \to N + \bar{\nu}. \qquad (1)$$

The first step, conversion of a muon into a pion and an anti-neutrino, is equivalent to the second reaction (43, text). Instead of a neutrino being added to the right-hand side of the equation, an anti-neutrino has been added to the left-hand side. The matrix element of the first step is obtained from (45, text) by substituting for w_s its order of magnitude μc^2. One finds

$$e_3 \hbar / \sqrt{2\Omega\mu}. \qquad (2)$$

The second step of (1) is a Yukawa process (38b, text). The order of magnitude of the matrix element is given by (42, text), where again μc^2 will be substituted for w_s. One finds

$$e_2 \hbar / \sqrt{2\Omega\mu}. \qquad (3)$$

The intermediate state of (1) lies energetically above the initial state. The order of magnitude of the energy difference is $(\mu - \mu_1)c^2$ which is the difference in the mass energies of the pion and the muon. According to (61, text) the apparent matrix element for the transition from the initial to the final state has the order of magnitude

$$\frac{e_2 e_3 \hbar^2}{2\Omega\mu(\mu - \mu_1)c^2}. \qquad (4)$$

The transition (1) is essentially equal to (52, text) from which it differs only because an anti-neutrino, instead of a neutrino, appears in (1). The matrix element for this transition has been given in (54, text) as g_3/Ω. By comparing this expression with (4) one finds

$$g_3 = \frac{e_2\, e_3\, \hbar^2}{2\mu(\mu - \mu_1)c^2} \tag{5}$$

which establishes a possible relationship between e_2, e_3, g_3. With the values $e_2 = 10^{-8}$, $e_3 = 1.2 \times 10^{-15}$ one calculates from (5), $g_3 = 4.9 \times 10^{-49}$, a value not very different from the one 1.3×10^{-49} found in Section 16.

The above argument could also be reversed. One might assume the Yukawa interaction and the interaction (52, text) to be primary and derive in a similar manner the existence and the strength of an interaction like (43, text), by the following two steps:

$$\Pi^- \to N + \bar{P} \to \mu^- + \bar{\nu}. \tag{6}$$

The first step is a modified Yukawa process (38b, text) in which instead of a proton on the right-hand side of the equation an anti-proton has been written on the left-hand side. The second step is obtained with similar changes from (52, text). From (6) one derives a relationship between the interaction constants by applying (61, text) and handling the sum, which is divergent, by the same questionable cut-off procedure adopted in Section 24. One finds

$$e_3 = \frac{e_2\, g_3\, M^2 c}{12\pi^2\, \hbar^3}. \tag{7}$$

With $e_2 = 10^{-8}$, $g_3 = 1.3 \times 10^{-49}$, this formula yields

$$e_3 = 8 \times 10^{-16}$$

which differs by only a factor 1.5 from the value of e_3 found in Section 13.

The following difficulty should be mentioned. If the previous analysis of process (6) is correct, one might expect that a similar process

$$\Pi^- \to N + \bar{P} \to e + \bar{\nu} \tag{8}$$

should also exist. The second step of (8) is like a beta-interaction step (46, text) in which the proton is replaced by an anti-proton on the opposite side of the equation. Formula (8) would lead to a spontaneous disintegration of the pion into an electron and a neutrino which should compete with the disintegration process (6) and is not observed experimentally.

If one takes literally the methods outlined in this appendix one would expect reaction (8) to be somewhat faster than (6) while experimentally it is slower by at least a factor 10 and perhaps does not happen at all. It appears, therefore, that some selection rule is operating which either prevents or slows up (8) appreciably.

Shortly after announcing his theory of nuclear forces Yukawa proposed a theoretical explanation of the beta interaction (46, text) as due to a process which would be described with our notations as follows:

$$N \rightarrow P + \Pi^- \rightarrow P + e + \bar{\nu}. \qquad (9)$$

It is assumed here that the pion could disintegrate into an electron and an anti-neutrino:

$$\Pi^- \rightarrow e + \bar{\nu}. \qquad (10)$$

The reaction (9) would therefore describe the disintegration, according to (10), of the virtual pion which is present part of the time in the vicinity of the neutron.

If process (10) should exist, the analysis of the two-step reaction (9) would establish a relationship between the beta-interaction constant g_1, the Yukawa constant e_2, and the rate of (10). It seems, however, that the experimental upper limit to the rate of (10) is too slow to permit an adequate explanation of the beta transitions on this basis.